FRAGILE OBJECTS

FRAGILE OBJECTS

Soft Matter, Hard Science,
and the Thrill of Discovery

Pierre-Gilles de Gennes
Jacques Badoz

Translated by Axel Reisinger

COPERNICUS
AN IMPRINT OF SPRINGER-VERLAG

Published in the United States by Copernicus, an imprint of Springer-Verlag New York, Inc.

Copernicus
Springer-Verlag New York, Inc.
175 Fifth Avenue
New York, NY 10010

Cover art: A soap film colored by the interference of light (see Chapter 7). Note the three black areas, where the film is too thin to reflect light; such films were used for divination by the Babylonians. (Photo courtesy K. Mysels.)
Translated from the French *Les Objets Fragiles*, by Pierre-Gilles de Gennes and Jacques Badoz, published by Librairie Plon, Paris, France, © 1994.

Library of Congress Cataloging-in-Publication Data

Gennes, Pierre-Gilles de.
[Objets fragiles. English]
Fragile objects/Pierre-Gilles de Gennes, Jacques Badoz; translated by Axel Reisinger.
p. cm.
Includes bibliographical references.
ISBN-13:978-1-4612-7528-2 e-ISBN-13: 978-1-4612-2410-5
DOI: 10.1007/978-1-4612-2410-5
1. Matter—Miscellanea. 2. Surfaces (Physics)—Miscellanea. 3. Research—Miscellanea. 4. Education—Miscellanea. I. Badoz, J. II. Title.
QC171.2.G4613 1996
530—dc20 96-17039

Manufactured in the United States of America
Printed on acid-free paper.

9 8 7 6 5 4 3 2 1

SPIN 10524616

PREFACE

School Stories

This book chronicles a journey through the high schools of France and overseas territories, all the way to the French Martinique.

I had long felt the need to speak to high school students. But that proved impossible. I happen to be the director of what is called in France, a bit pompously, an institute of higher learning (*une grande école*): the Institute of Physics and Chemistry, in Paris. Had I proposed to a high school principal to give a lecture to his students, he would have replied, with some justification, that it would constitute covert propaganda for my institute, and a visit was, therefore, inadvisable.

But, as fate would have it, the Nobel Committee elected to award me a prize. As a result, from the end of 1991 on, I found myself invited to speak at high schools: most often at the direct initiative of the students themselves, sometimes under the umbrella of student associations, science clubs, etc., but also at the urging of some progressive teachers.

So I took to the road, armed with a set of slides under my arm, to tell about the life of my researchers: what kinds of questions they ask themselves, what their life style is in their work. To share their successes, their failures, their disagreements, and their mistakes. To show that science is on the move, but also that its march is not a straight path, that it is filled with zigzags and traps, as well as unexpected shortcuts. To convey the message that the research scientist is neither a prophet nor an Einstein-like savant, but rather like an explorer, often hesitant and exhausted.

From the very first schools I visited, I experienced a feeling of utter joy—in spite of logistical snafus: auditoriums with barely adequate sound systems, balky slide projectors, projection screens no larger than postage stamps, all details, to be sure, but important ones, when one is faced with one or two thousand excitable adolescents in a distant suburb.

Utter joy because, in spite of everything, a special silence would take hold from the opening words. One could sense the electricity. The formal talk was followed by a question and answer period, with the real debate. It was invariably difficult to get off the ground (particularly if school officials were in attendance). But with a little prodding, the first embers would start glowing, and pretty soon you would find yourself at the center of a bonfire: one or two hours of uninterrupted questions!

The first questions were typically of a technical nature, concerning such and such property of rubber or glue. Then the scope of the inquiries would broaden:

- What path should one follow to embrace scientific research?
- Is there a unique type of research scientist?
- Does one have to be proficient in math?
- What is your assessment of the state of education in our high schools?

Gradually, the questions would begin to address truly general issues:

- Why go into the sciences?
- What are the possible dangers?
- Is it even desirable to fundamentally alter the world and our environment?

And eventually:

- What is our world?
- Where do we come from and where are we going?
- Are there any other life forms in the universe?
- Is there an afterlife?

It is an awesome challenge to have to face such questions, to which no one has satisfactory answers, but which must nonetheless be asked. (It seems that I have just defined the field of philosophy, a science to which I am largely a stranger.)

What a joy it is to reflect on such questions with an entire audience, to grope together for the truth.

Another aspect, predictable enough, of these discussions concerns the topic of education. From the very beginning of my talk, I would often fire critical barbs at the weaknesses of the French system. And the reaction was immediate, taking on the attributes of a triangular relationship of sorts between the students (who are in the thick of the game), the teachers (who follow the action from the sideline), and me (who tries to return the ball into play as best as I can).

Quite often, my visits took off in yet other directions. A particular class, for instance, had prepared a great experiment; another had set up a science show, or a workshop. It would provide the opportunity to speak more quietly to smaller groups, in a laboratory, on a lawn, or even while enjoying a few drinks.

There are so many accumulated memories—from the large traditional high schools to the small professional schools (often the liveliest ones). I think, for instance, about Rochefort, where young people are taught how to manufacture plastics. There was in Rochefort no meeting hall large enough to accommodate my presentation. So I talked in the machine shop, with all the students gathered on one side, the technicians in their blue lab coats on the other, in an atmosphere that reminded me of Jules Verne.

Nor will I ever forget the high school in Asnières, where the most suitable meeting hall was the theater, made available by an enthusiastic principal. The place was packed to the point that I, who move about a great deal while I talk, had to step over bodies overflowing every inch of the stage.

There were many different kinds of theaters in this story, from the formal and charming halls in Nancy or Besançon, to restored movie houses, even to athletic fields like the famous stadium in Orthez or the Lamentin stadium in the Martinique. How could I forget the bold, incisive questions asked by the children of the Antilles, or the moving dedication of a training laboratory, with the ceremony presided over a 16-year-old student?

I have accumulated a veritable harvest of memories. But eventually, I had to stop: I was overloaded, exhausted. With pain in my heart, I was forced to turn down every invitation for a year. I had to reply "no" to over 150 active, enthusiastic, and creative high school organizations. It was impossible to close this happy page in my life without trying to capture its memories. From this desire sprang the idea of chronicling the story in a book—a somewhat naive plan, as it turned out. There was plenty of material available: audio tapes, photographs, videotapes (which focus on the speaker but fail to capture the feedback on which I rely so constantly). My good fortune was to find along the way a friend, Jacques Badoz, himself a teacher at the Institute of Physics and Chemistry. He became interested in these trips and, more particularly, in the discussions they generated. He has collected and repackaged all the material. He has spotted the most glaring shortcomings, and of this chaotic mixture, he has created a coherent whole. His vision was very close to mine: we worked together swiftly and painlessly.

At this point, it is incumbent on me to pause and warn my readers, gentlemen and fair ladies (as once did the author of a much greater Sentimental Journey), about what *they will not find* in these pages.

The questions posed by the students elicited from me answers which came from the heart, given the limited resources available at the time, and flavored with my own biases. We have endeavored to preserve the naiveté of these answers. For instance, when asked about a particular environmental problem, I merely express a point of view. I do not undertake a detailed analysis of the issue, which would require an entire book. Anybody wishing to form an educated opinion on a topic of this magnitude is urged to consult recent books on the subject, such as those by Zaher Massoud, Claude Allègre, or (the latest) Gérard Lambert.

In short, do not expect to find here figures, statistics, or detailed data on current problems. This is a book of memories, and nothing more.

The memory itself is fragile. We were unable to preserve the actual individual questions which, for the most part, were not recorded.

We had to recreate them, and even reshape responses on a particular topic given to several different audiences. In the process, we have undoubtedly lost much of the spontaneous nature of the dialogue. But, to the extent possible, we wanted to capture the nascent sparks of curiosity and enthusiasm which were so pervasive throughout the schools we visited.

Paris, France | Pierre-Gilles de Gennes

ACKNOWLEDGMENTS

This book is based on audio or videotapes made during P.-G. deGennes's visits to various organizations.

We are very grateful to those—amateurs, video clubs, or professionals—who produced them and who so generously loaned them to us (sometimes for an extensive period). We are also indebted to all those who made them possible: the principals, teachers, documentary producers, and students of the following schools.

- Lycée Alain—Le Vésinet
- Lycée d'Arsonval—Saint-Maur
- Lycée Guez-de-Balzac and its video club—Angoulême
- Lycée Claude Bernard—Paris
- Lycée Chateaubriand—Rennes
- Lycée Marcel-Dassault—Rochefort
- Lycée L.R.-Duschesnes—La-Celle-Saint-Cloud
- Lycée Jules-Fil—Carcassone
- Lycée Fulbert—Chartres
- Lycées de Guadeloupe (convention on the Fouillole campus)
- Lycées and Colleges of La Martinique (convention at Le Lamentin and on the Schoelcher campus)
- Lycée Notre-Dame—St-Germain-en-Laye
- Lycée Notre-Dame-de-Bury—Margency
- Lycée "Les orphelins et apprentis d'Auteuil"—Paris
- Lycée Les Pierres-Vives—Carrières-sur-Seine
- Lycée de la Planoise—Besançon
- Lycée Raynouard (and City Hall's communications office)—Brignoles
- Lycée Auguste Renoir—Asnières

We are likewise indebted to various organizations which de Gennes addressed:

- Commission du Titre de l'Ingénieur
- École Européenne d'Ingénieurs en Génie des Matériaux (Institut Polytechnique de Lorraine)—Nancy
- École Technique Supérieure du Laboratoire—Paris
- Institut Universitaire de Technologie—Angers
- Physis-Orbis, Association des Étudiants de l'École de Physique—Magistère de Grenoble (Création DS Vidéo Communication)—Grenoble

Also used was the material presented during three productive meetings with teachers' organizations:

- Instituteurs du 19e Arrondissement, Paris
- Enseignants de Science en Guadeloupe
- Instituteurs au Lamentin en Martinique

We are deeply grateful to organizers and audiences alike.

Paris, France — Pierre-Gilles de Gennes
Paris, France — Jacques Badoz

CONTENTS

PART II * Research

PART I

Soft Matter

CHAPTER 1

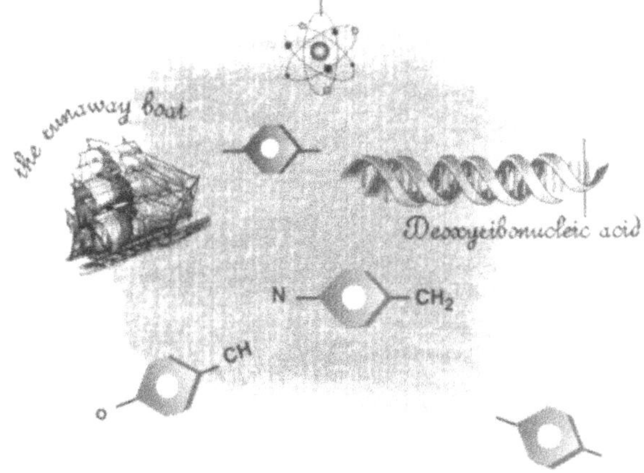

The Indian Boot and Mr. Goodyear

In discussing long-chained molecules, also called *polymers* [1],[1] I like to start with a story based in the Amazon. Indians of the Amazon take the sap from the hevea tree—a sort of whitish, milky juice like that of the common dandelion—and smear their feet with it. At first, this sap is like an ordinary liquid: it flows. But after about 20 minutes, it coagulates, and the Indian has made himself a boot (Figure 1).

This is a most interesting transformation. I will start by explaining how we understand it today, 2500 years after the Indians discovered the trick.

The initial liquid, or *latex*, contains long-chained *molecules* [2], which can be visualized somewhat like spaghetti in a soup, quite soft and quite flexible. If one shrinks the size of the noodles we normally eat by a factor of 100,000, one gets a fairly accurate picture of the structure of this latex.

---

[1] Numbers in brackets refer to entries in the glossary at the end of the book.

Figure 1

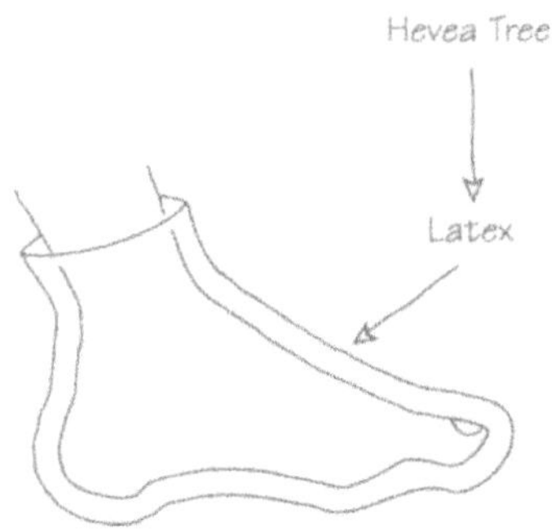

A "boot" invented 2500 years ago by South American Indians.

After the Indian has coated his foot, a new actor comes into play: oxygen in the air. Oxygen causes a remarkable reaction: it ties the chains to one another at a few spots (Figure 2). If one has a dish of spaghetti, it is easy to suck on a single strand and slurp it in. Stick the noodles together at a few points with drops of glue, and that is no longer possible. If one pulls on one of them, the entire network resists: *The contents of the dish have gone through a transition from a liquid to a solid.*

Yet the solid is somewhat unusual. If you could navigate through that structure like a miniature fish the size of a few *atoms* [3], you would notice no difference before and after the oxygen had reacted. Everything would still move more or less the same way as before and it would still look like a fluid. So, the substance used as a boot by the Indian is solid on a macroscopic scale, but fluid on a microscopic one. This is what we call *rubber.* (Incidentally, the miniature fish roaming about the structure which I was just talking about is not a purely imaginary concept. It exists in the form of tiny magnets carried by certain atomic nuclei. A technique called *nuclear magnetic resonance* can track the signals emitted by these tiny magnets and allows us to probe the immediate surroundings of a molecule on a scale of a few angstroms [4]. The measurements confirm that, on a local scale, rubber is indeed still a liquid.

As ingenious as the Indian boot may seem, it is not very satisfactory: it disintegrates after only about a day. Here is why: The oxygen from

Figure 2

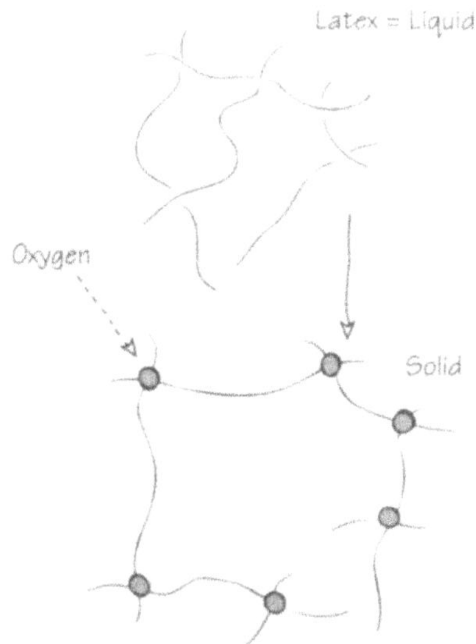

Changes in the structure of latex under the influence of oxygen.

the air at first produces a useful reaction when it ties the chains together. However, because it is fairly reactive, its chemical action on the chains continues, and in a second stage, the oxygen eventually cuts the chains themselves. It is as though we were to slash haphazardly with a pair of scissors the links of a fisherman's net. At some point, the net will fall apart. That is precisely what happens to the Indian boot: it ends up losing its mechanical integrity and simply breaks down.

How can this be prevented? With a process invented in 1839 by an American whose name is familiar to everybody: Goodyear.

Small Cause, Large Effects

The year is 1839. It is a time of dizzying developments in chemistry. Chemists are just starting to synthesize innumerable new molecules. They try a little bit of everything, they react everything with everything else. One among them, Charles Goodyear (1800–60), prompted by sheer curiosity, takes the latex from the hevea tree and decides to boil it with sulfur. He has no idea what latex is. He only knows that it contains carbon and hydrogen. The very concept of a long-chained molecule is totally unknown to him. But he continues to experiment . . . and succeeds in moving things forward. He obtains a sort of blackish substance, deformable and yet resilient: *natural rubber.* This rubber remains, a century and a half after Goodyear, one of the pillars of man's industrial activity. The ordinary tire of an automobile is composed of 16 elastic layers, each with a different function. Of these 16 layers, several are made, to this day, of Goodyear's natural rubber.

No one really knows why Goodyear was inspired with the idea of using sulfur, but we understand today the key to his success. Sulfur has chemical properties fairly similar to those of oxygen—it is in the same column of the *periodic table of elements* [7]—but it is slightly less reactive. It is still capable of selectively tying the chains together in a few spots. However, unlike oxygen, it is not potent enough to cut the chains themselves. And that is the secret behind the stability of natural rubber.

One last remark about this product: if one counts the number of carbon atoms (C) that have reacted with the sulfur, one finds typically that only 1 in 200 has done so. Yet this extremely weak chemical reaction is enough to cause a change in physical state as profound as a transition from a liquid to a solid: fluid to rubber. This proves that matter can be transformed by weak external actions, much as a sculptor deforms the shape of clay by the gentle pressure of a thumb. This is the central and fundamental definition of *soft matter.*

The story of the invention of rubber is not the kind one likes to tell in the presence of decision makers in ministries and government agencies. For their benefit, we prefer a different version, something along the following lines: "In order to manufacture novel materials, one must first understand the structure of matter, which implies a considerable amount of research. Only then will we be able to manufacture the useful material, with the required properties to meet the needs of a particular application." This story has been valid in certain cases (such as semiconductors, or the laser) [5]. But in other cases, like that of natural rubber, the practical invention came long before any understanding of the physical phenomena involved.

CHAPTER 2

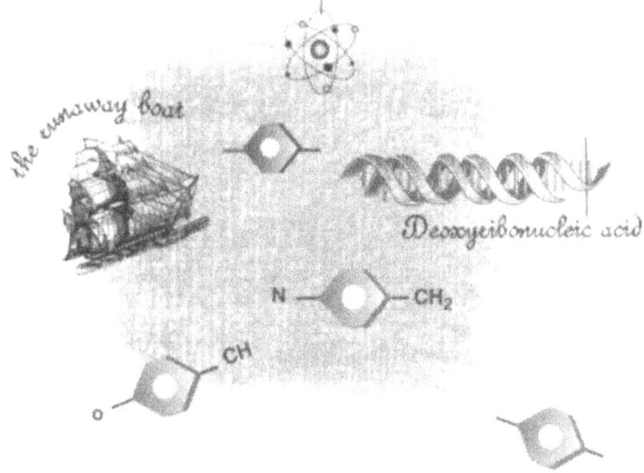

The Noodle Soup

Let's return to our natural rubber. As I said earlier, Goodyear had, in 1839, no idea about the structure of his miracle product, nor did he or any other chemist know anything about the existence of long-chained molecules—polymers or macromolecules. Indeed, no chemist had any interest in this type of product, and with few exceptions, when some was made inadvertently, it was immediately dumped into the sink. A grave error, which caused a 50-year delay!

Why this attitude? When a nineteenth-century chemist tried to synthesize a new material, he attempted to react two known compounds on each other. He probably did not know just what the result of his creation was going to be—that is precisely what he was trying to find out—but he expected to obtain a molecule containing at most a few tens of atoms (about the same number as in the starting materials). He was not prepared to find a compound whose every molecule contains a million atoms! Besides, unlike today's chemist, he lacked the tools to determine whether he had made one product or a mixture of several. The standard practice then was to purify the product of the reaction by crystallization or sequential distillation until one obtained a pure substance. The main criterion for purity was the melting point. If the compound obtained is pure, it has a well-defined melting point: when it is heated up, it changes from a solid state to a liquid state at a very precise temperature. If it is a mixture, the melting process starts and ends over a range of temperatures.

With this criterion in mind, our chemist would select pure materials based on properties that he could study with the tools at his disposal. As it happens, long-chained molecules have (in general) no well-defined melting point. They do not crystallize unambiguously and, when heated up, they behave somewhat like amorphous glass. The solid starts softening before liquefying, and the transition from solid to liquid takes place over a rather extended temperature range.

Given the purity criteria of nineteenth-century chemistry, these molecules were doomed, and the chemists of the time had no reason to show any interest in them.

A lesson can be drawn from this historical sketch. We often strive in our classrooms to describe phenomena in physics and chemistry in terms of laws, doctrines, and rules. The advantage of this approach is that it allows us to summarize a lot of knowledge in a compact framework. But it remains essential to preserve one's freedom, to maintain one's perspective, and, at times, to keep one's ability to deal with novel questions. For instance, to ask: "This product I have just synthesized, or this phenomenon I have just observed, appears not to conform to current knowledge. Is it the result of a trivial experimental error, are we justified in ignoring it, or should we examine it from a fresh angle in order to understand something fundamentally new?"

But let's go back to polymers. It was not until 1920 that the concept of long-chained molecules, or macromolecules, was accepted by the community of chemists and physicists. It happened thanks to a great chemist named Hermann Staudinger (1881–1965), who succeeded around 1920 in a delicate experiment. He synthesized relatively long molecules (some 20 units, rather than 1000 or 100,000 units). Chains of this length are still short enough to be crystallized and to have a well-defined melting point, so they appear like well-characterized, pure substances. Their molecular mass can be determined.

From that point on, gradually, by means of successive operations, Staudinger prepared ever longer chains. By virtue of their very method of synthesis—by continuity, as it were—these chains remained well-defined, even though their melting point becomes increasingly fuzzy. From all this, he concluded that it is possible to make very long molecules. And this realization ushered in the beginning of a new era. A full 80 years had elapsed since Goodyear's natural rubber.

But even Staudinger was a bit too much the prisoner of the doctrines of his time. While he had established clearly the chemical sequence involved in making certain polymers, he had an erroneous concept of what these chains look like in space. He believed that each of the macromolecules that he had made was a small, rigid rod (about a *micron* in length) [4].

It was another 20 years before another great German researcher, Richard Kuhn (1900–67), successfully demonstrated that these macromolecules are not stiff rods at all, but are, instead, quite flexible objects (our noodle analogy). This idea, in turn, allowed Kuhn to explain the amazing elasticity of rubber.

Kuhn was a remarkable individual. Before getting interested in macromolecules, he was, in the years between 1920 and 1930, among the pioneers of the new "quantum mechanics," which describes the behavior of electrons in atoms and molecules. (We owe him, by the way, one of the great spectroscopic rules, one which bears his name.) But in the 1930s, in spite of his success in the area of atomic physics, Kuhn decided to switch fields. He settled upon the physical chemistry of macromolecules. To the study of these objects—which were considered somewhat bizarre, a bit "unclean"—he brought an extraordinary contribution: the understanding of their flexibility, and what conclusions to draw from that understanding.

Questions and Answers

This account, however sketchy, of the discovery of macromolecules, has elicited a number of questions, mostly about the relationship between *fundamental* and *applied* research. Here is a story which starts with a purely empirical, pragmatic discovery: that of rubber. It is followed later by a lot of systematic research:

1. To understand, first of all, the nature of these long molecules.
2. To learn how to synthesize them from small chemical sequences—the elementary building blocks—which are repeated sequentially to create these molecules.
3. To understand their properties, for example, the mechanical characteristics of a particular plastic.

The story leads to a harvest of practical results: few of us are aware of the number of objects surrounding us, of the textiles we wear, which are manufactured from materials made up of long-chain molecules. Even fewer are aware that we, ourselves, are made of the very same type of molecules.

There is also the question raised by Kuhn's career path. When, how, and why, might one wonder, should or can a research scientist change his specialty? "Why," I have often been asked, "did you switch from the study of superconductors to that of soft materials?"

It would be dangerous, in attempting to answer such questions, to insist on formulating rigid rules. History and circumstances must influence our purpose. Thus I prefer to answer by quoting specific examples, anecdotes based on everyday life in research.

Applied Research and Fundamental Research

Let's start with this odd couple, so often cast as adversaries.

Fundamental Research, a Necessary Responsibility of Developed Nations

To find out how the world functions, from the atom to the galaxies, from minerals to living beings, is an exalting quest. This progress of knowledge is independent of any practical result that might ultimately be derived from such an effort. The journey of the *Voyager* spacecraft to the edge of the solar system needs no other justification than the harvest of knowledge that rewarded us. The discovery of the double helix structure of the genetic code would have been satisfaction enough, even without the phenomenal practical possibilities that we envision today.

Trapping a single atom in a cell and "cooling" it with laser beams down to one millionth of a kelvin is another marvelous experiment, just mastered recently, and (so far) devoid of any practical use.

One can sometimes justify fundamental research by projecting the potential, hidden applications that it might spawn. The study of the electronic properties of solids was pregnant with the invention of the transistor, the fabulous development of microelectronics, the advent of digital technology. Spectroscopy gave birth to the invention of the laser, which revolutionized communications.

But apart from of any potential payoff, I believe that the quest for knowledge for the sake of knowledge is irreplaceable. It seems to me essential that a developed nation participate in this expansion of knowledge for the good of all mankind, and that it devote to this effort a fraction of its resources, even without any firm hope of material benefit in return. If unexpected applications do materialize as a bonus, so much the better.

A few remarks are in order, however. This "pure" science does have an obligation. It must produce truly original results. There is no place for research which merely reproduces, except for a detail or two, an experiment already successfully performed. But how does one judge the degree of novelty? At present, this difficult evaluation is carried out informally, by the seat of the pants so to speak, by the scientific community—not entirely without errors or injustices, to be sure, but, by and large, with fairly satisfactory results.

A word of caution: it does sometimes happen that, in order to reach a particular goal, several avenues prove feasible, utilizing very different means, some quite elaborate, others much simpler or considerably less costly. Should one, for instance, send humans into space, or should one rely on unmanned spacecraft? One must constantly remain alert to the danger that research might get mired in enormous projects of doubtful utility. This is the responsibility that scientists must assume on behalf of the taxpayers of their respective countries.

Applied Research: A School for the Imagination

Every piece of applied research starts from a challenge: to develop a new product, to fill a new need that is not being met in the present system, to offer people better lives. And this challenge must be answered by utilizing the entire gamut of knowledge, from the most fundamental (for instance, the properties of magnetic films on interactive video disks) to the most practical (how to get the ink in a ball point pen to flow properly). It is the very richness of this palette, the varied and unexpected choice of colors, that makes for the greatness of applied research.

In my opinion, the merits of a great inventor are no less admirable than those of a great theorist. I might even have more respect for the inventor. In the nineteenth century, it was typically a lone individual who waged a battle (often on several fronts) with incredible tenacity. In the twentieth century, new products are most often developed by industrial teams, better equipped perhaps, but pressured just the same by a fierce competition.

My sense is that if an extraterrestrial were to peek at us and evaluate us, he would note that man was bold enough to have imagined black holes, but that he was also clever enough to invent the zipper.

The Marriage of the Fundamental and the Applied

I must emphasize the mutual advantages resulting from the interplay between fundamental and applied research. As our small team at the Collège de France was beginning to investigate polymers, we were intrigued by the motion of these large molecules in solution, a matter of pure curiosity for a fundamentalist. Subsequently, we were exposed to more and more problems of various types as a result of our dialogues with industry. One afternoon, while visiting Allied Chemical in New Jersey, we were told about a peculiar problem. The dissolution of some polymers in solvents is difficult, and the technologists did not clearly understand what parameters were available to accelerate the process. It was clear that finer powders dissolve more easily, but nobody knew just how finely the powder had to be ground. With the naiveté characteristic of a theoretician, I instantly proposed an explanation, and blurted out "I believe that this is a problem of cooperative diffusion, and that such and such rule is applicable." No, they replied patiently, we thought about it, but experiments show conclusively that it is much more subtle than that."

After returning to Paris, we discussed this problem with my colleague Françoise Brochard during the ensuing months. We finally hit upon an absolutely fascinating mechanism which effectively superposes two distinct processes and introduces a new characteristic length of the polymer. We called it the "magic length." It defines the size down to which the powder must be reduced in order to ensure the best possible dissolution. This dimension happens to be much larger than that of the macromolecules themselves. It serves no useful purpose to reduce the size of the powder below the magic length. Here is the case of an initially very practical problem that led us down a path of fundamental study of the dynamics of intertwined polymers.

This type of situation comes up fairly frequently in our field of research. One must have the ability to listen with some humility to the individual in the trenches who tells you "here is something peculiar," then ponder over it and digest it. In a few fortunate cases, we find a solution. Most of the time, we fail. But on a global scale, our position as fundamental research scientists is a rather pleasant one in that we get to choose our problems for the pleasure of understanding them, and once in a while we see our efforts rewarded by a (modest) practical payoff.

Working on problems related to concrete applications also provides a degree of reassurance in trying to recruit young student theoreticians. At the completion of their thesis, they have the option of going into fundamental research. But they are equally comfortable choosing careers as research engineers, dealing with issues of immediate applicability. One of my proudest accomplishments is to have among my former students one of the research directors at Philips.

Here is another example of a problem with both applied and fundamental aspects. The Collège de France team became interested in the problem of recovering oil after a field has nominally been depleted, although a substantial amount of oil remains trapped in the porous rocks unable to flow to the pumps. We studied, among other techniques, the process of foam-assisted oil recovery. Briefly, foam is injected through one well, and if everything works out according to plan, oil is squeezed out of a neighboring well. As peculiar as the technique may sound, it offers some hope of recovering at least part of the residual oil. But this very practical problem has a fundamental aspect as well, for understanding the motion of a foam through a porous medium is a difficult problem of statistical physics. It is not unreasonable to expect that students working on this fundamental physics problem may at some later time find a job in the oil industry (assuming that the price of a barrel goes up).

Kuhn's Example: Knowing When to Switch

"A rolling stone gathers no moss." That piece of "wisdom" in virtually every culture does not seem to support Kuhn's decision. As I mentioned earlier, he gave up the promising young field of atomic physics to embark on the ostensibly very risky study of polymers and macromolecules.

Well Then, Ready for a Change?

Everybody's scientific career is an individual case. But it might be useful, nevertheless, to recall here some specific examples.

In the years 1961–1965, I was interested in superconducting metals [5]. These are truly amazing metals, which, at very low temperatures, carry an electric current without a trace of loss. Lead, tin, and mercury are classic examples. The phenomenon had been known since 1911. Around 1957, thanks to the brilliant insight of a young student named Leon Cooper, some understanding of how this works was beginning to take shape. We got into the game in 1961. At first, we worked with materials that were easy to prepare, like lead-tin alloys. As we gained experience, we wanted to try other materials with better performance, such as niobium-tin. This alloy is quite fragile, difficult to draw into wires, and generally poses enormous metallurgical problems. The challenge is to control the small precipitates that play a role in "pinning" the magnetic flux, that is, in preventing magnetic fields to roam freely through the material, disrupting the flow of electric currents. This kind of delicate metallurgy requires heavy and costly hardware, such as large electron microscopes. At this stage, one is faced with two choices: either one becomes a genuine metallurgist and requests massive resources to create an adequate laboratory, or one remains a tinkerer dabbling in "light" science, destined to make a discreet exit. I chose the latter option.

Another aspect also came into play. My co-workers who wished to continue in this field had become sufficiently proficient to work without me. Had I stayed in, I probably would have been more of a hindrance than a help.

Time to Move on, But Where To?

In the early part of the century, in the 1920s, the study of liquid crystals had made remarkable strides in France under the leadership of George Friedel. With common materials and a simple microscope, through observations on an ordinary scale, he was able to establish the structure of these materials at the molecular level. A fabulous accomplishment indeed. Later on, in the 1930s and 1940s, physicists in the Soviet Union revisited the issue. They too contributed spectacular advances. But the Soviets suffer a little bit from the same flaw that afflicts us in France: they have too much faith in theory. They get interested in idealized objects, far removed from real molecules. They dug deep into the mine, but failed to notice the nuggets within easy reach near the surface. We suspected the existence of a rich vein around 1968. This time around, we did not have to create a new group. We convinced already established groups to join with us to work on liquid crystals. Within three years, the Orsay team had staked out a position of world-class status.

Five or six years later, the topic had become more technical, more industrial, and required more formidable resources. I began to think that it was no longer my calling.

At about the same time (toward 1972), many questions concerning polymers surfaced. Kuhn's description of long, flexible chains, like noodles in a soup, represented significant progress. Nevertheless, there remained many statistical problems to solve in order to understand how these chains arrange themselves in solution. Two chains repel each other. Within a given chain, one fragment repels another. All this gives rise to a situation that was difficult to analyze. Fortunately, we had several trump cards. Our background was rooted in our expertise in the areas of phase transitions and critical phenomena. When water is changing to steam, there is a certain *critical point* of pressure and temperature when the liquid and its vapor have almost the same density, indeed are nearly identical. Under these conditions one observes extraordinary phenomena, such as microdrops, drops within drops, and so forth. We were fortunate to find a way to take concepts developed in the framework of critical points and to adapt them to the seemingly very different field of polymer chains floating in a solvent (the noodles in the soup).

The Dish of Noodles

Incidentally, I cannot emphasize enough the importance of this transposition of methods between two apparently unrelated fields of science. What has been learned in one field can at times help solve completely different problems. This offers a considerable savings of effort: once the link between polymers and critical phenomena was established, we suddenly had at our disposal, in a single stroke, 20 years' worth of sophisticated knowledge contributed by the study of critical phenomena. We did not have to start from scratch!

We were also lucky that a young team (that of G. Jannink), which worked with neutrons at the Atomic Energy Commission, developed an interest in polymers at the same time. Neutrons turn out to be a powerful tool for studying macromolecules. Finally, there was in Strasbourg a first-rate laboratory—the CRM, French acronym for Center for Macromolecule Research—founded by a great physical chemist, Charles Sadron. They welcomed us with open arms, and in the process spared us a lot of stupid mistakes.

From the behavior of macromolecules among themselves, we moved quite naturally to the problem of their behavior toward other substances, notably solid grains in suspension, which led us to the science of colloids, or suspensions of fine particles. More about those later.

"To Leave Is to Die a Little" (E. Haraucourt)

I have just described the chronological evolution of our group, perhaps in a way that is too utopian. In truth, switching fields is not as easy as I made it sound, even for a theorist. It means some three years of work, each time, just to come up to speed.

For an experimentalist, the prospect is even more daunting. He also has to contend with a traumatic change in instrumental expertise, a completely new set of scientific apparatus. To change disciplines is to leave one scientific community for another. Often, these communities have different languages, different books, different instruments. Just about everything is different. Still, I am happy to have participated and helped in two or three such changes.

CHAPTER 3

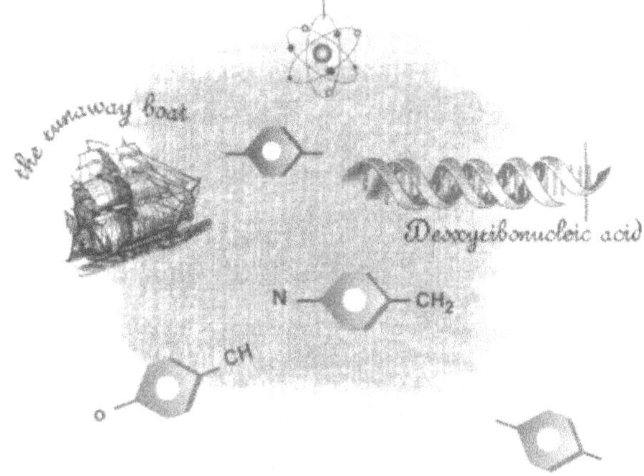

The Tubeless Siphon and the Runaway Boat

Large molecules have played an important part in the development of life from its beginning. For a long time, humans were content to use, without any effort to transform them, the many examples provided by nature: the long cellulose molecules of cotton, the polymeric oils of linseed; the long molecules of silk, already more or less aligned, forming natural fibers. And we mustn't forget the structurally important cellulose of wood, the keratin in the wool of the nomadic shepherd, the cellulose in the papyrus of the Egyptian scribe, and others.

Very early on, however, some transformations appear in the history of technology, involving empirically discovered chemical reactions. I have already mentioned the formation of rubber in the Indian's boot. The Renaissance painter used linseed oil to advantage. He waited for his painting to "dry," unaware that oxygen in the air attaches long molecules contained in linseed oil at a multitude of points (in contrast to the Indian's rubber), until it turns it into a solid, hard and resistant. When the painter adds a *siccative*—a drying agent, which is an oxidizer—he need not wait quite as long. The oxidizer speeds up the action of oxygen in the air and, hence, decreases the drying time.

The hairdresser who gives a lady a permanent may not realize that the curling iron's heat creates links (because of the sulfur atoms contained in the hair's keratin) that "freeze" the curls. The dyer who attaches tiny coloring molecules to the long chains of cotton is making use of a more complex transformation. He is relying on a genuine chemical reaction, albeit one that is only partially successful. The weak bonding of indigo dye to cotton fibers created the legend of the "blue men" of the Sahara: their headscarves stained their skin.

A Construction Set

Around 1930, Staudinger's work opened two avenues, which have been vigorously pursued by chemists.

The first is the analysis of the chemical structure of natural macromolecules. This effort leads to the determination of the chemical patterns—generally of rather simple molecules—which, by repeated action, form a large molecule. I will mention two of the most notable successes:

- The determination of the structure of silk
- The more recent elucidation of the structure of deoxyribonucleic acid (DNA). DNA contains the genetic blueprint for each individual. This information is embodied in a sequence of four nucleotides (four small molecules) throughout the length of the DNA chain—an alphabet of four letters, as it were.

The second avenue pursued by chemists is the synthesis of entirely new macromolecules. Consider the magnificent construction set which is available to them. They need only help themselves from the catalog of innumerable organic molecules prepared and carefully described by chemists between the years 1830 and 1930. There is more than enough there to build, practically at will, a molecule which provides an extremely strong fiber (*nylon*, similar to synthetic silk), a light and resistant material (*kevlar*, used in bulletproof vests), or the variety of easy-to-mold plastics that have become such an integral part of our domestic appliances.

The Tubeless Siphon

Macromolecules that can form solids have properties of astounding diversity. But I want to focus here on some other properties, no less remarkable, of polymers which have a dramatic effect when added in minute quantities to a liquid. Let us start with a mishap experienced by a careless motorist: running out of gas. Assume that he manages to find a gas station and to fill his emergency can. But this is not his lucky day. The opening of the gas tank is such that he cannot pour in the precious fuel! His only recourse is to find a tube and siphon out the content of the can: he sucks on the gasoline, filling the tube, and without letting it drain out, shoves its extremity into the tank. Most importantly, he must position the can above the level of the tank (Figure 3).

If during the process the end of the tube (at point *B* in Figure 3) pops out of the gasoline in the can (a 1-millimeter gap is enough), air bubbles rush into the tube and "unprime" the siphon. He has to start all over again—a very unpleasant business.

About 15 years ago, my colleague D.F. James at the University of Toronto, Canada, performed a stunning experiment dealing with this phenomenon. Since a siphon runs dry under similar conditions with either water or pure gasoline, he chose to work with water, which is a lot

Figure 3

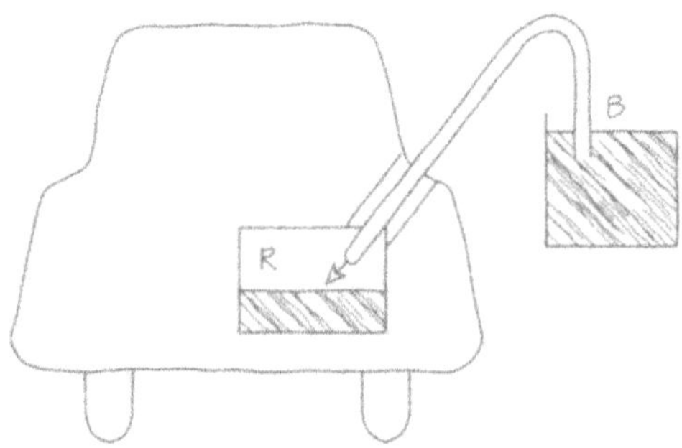

The ordinary siphon.

easier and less dangerous to handle in the laboratory. He dissolved a mere 1 gram of a polymer in 10 liters of water (1 part in 10,000). The polymer he chose was polyoxyethelene (polyox for short). With this tiny amount of polyox in the water, the siphon continues to work even with the end of the tube raised as much as 20 centimeters above the surface of the water. One actually sees a column of water rise into the air without any support! This phenomenon gave rise to the name tubeless siphon (Figure 4).

Here is a simplified explanation of what happens.

The polyox molecules dissolved in still water crumple onto themselves, but when caught in the flow of water, they unwind, exerting on the water a force analogous to that of a stretched spring (Figure 5). Those of you who do morning calisthenics with a bungee cord exerciser are familiar with this force, which is proportional to the elongation of the spring. In the liquid flow, the stretched polyox molecules "pull" on the water and, in doing so, compensate for the weight of the water column.

Figure 4

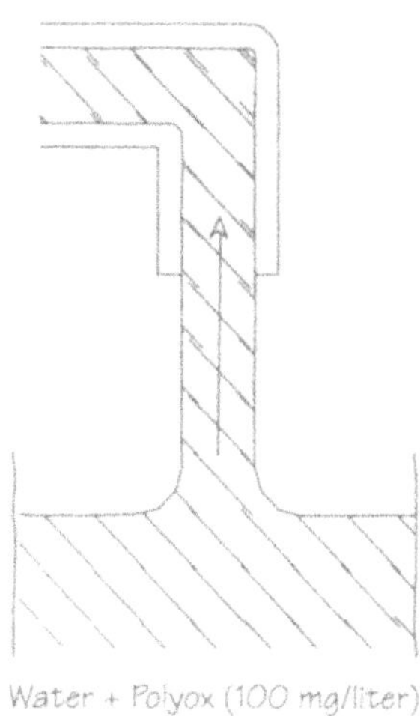

The tubeless siphon.

Figure 5

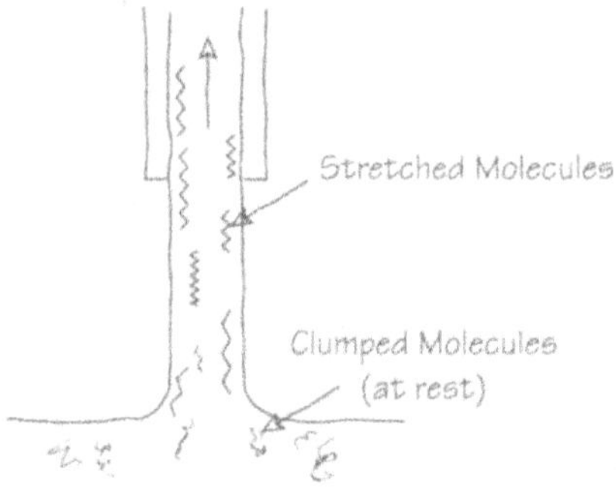

The physics of the tubeless siphon.

The Runaway Boat

Polymers in solution have other miraculous properties.

A fire breaks out in a high-rise building. Firemen aim their hoses (Figure 6). Despite the powerful pumps, the water jet reaches only the eighth floor. How can the young lady on the eleventh be rescued? A pinch of our miracle polymer—polyox—added to the water, and the height of the jet instantly increases by about 30 percent. She is saved! Here again, the quantity of polyox required is quite small (2 grams per 10 liters of water), and it is cheap to boot (only a few dollars per pound).

The city of Bristol, in Great Britain, found itself forced to improve its increasingly inadequate sewer system, built during the reign of Queen Victoria. A total overhaul of the network would have been tremendously expensive. A bit of polyox was enough to increase the flow in the pipes, sparing (at least temporarily) the pocketbooks of the taxpayers.

Polymers can sometimes play tricks. There is a shipyard in Paris, where the staff of the Navy Corps of Engineers uses an artificial body of water to test model boats, not for amusement, but to verify the calculations and projections of the engineering department before construction of a new type of ship. The model might be, for instance, 2 meters in size for a ship whose overall length is intended to be 200 meters (1/100th scale). One can predict, more or less empirically, the perfor-

Figure 6

Polyox saves the damsel in distress.

mance of the future ship from the measured characteristics of the model, for instance, its top speed for a given engine power. As good scientists, the engineers like to be sure of their results. So they repeat their measurements every few months under similar conditions: same model boat, same engine power, and same water in the pool. One day, panic set in: the results were no longer reproducible; remeasured several months later, the speed of a model boat had shot up by 40 percent—way beyond any possible experimental error. Uncertainties of that magnitude are unacceptable. The perplexed engineers finally discovered the culprit: it was the water in the tank. This water, which would be too costly to change frequently, remains stagnant for months at a time in the huge pool. These conditions are conducive to the growth of small algae which secrete minute amounts of certain polymers (in this case, polysaccharides, which are long-chained sugars). As it turns out, these chains have the same propensity to reduce friction as does the polyox in the sewers of Bristol! Armed with this observation, our engineers quickly solved the problem: they added a little more chlorine to the pool to prevent the growth of algae. The speed of the model boat promptly dropped back down to a more sedate and normal pace.

But the fundamental question remains: Why, in all these cases, do small quantities of polymers in solution reduce the friction between the fluid and a solid body?

The Limitations of Theory

The observed reduction of the drag—another term for friction—when a fluid flows in a pipe, or when a solid surface moves in a fluid (like the hull of the boat in the pool), is quite troubling. One might think intuitively that adding large molecules to water (our noodles in the soup) would increase the viscosity of the liquid. A viscous liquid flows less readily than a more mobile one. Even in the land of Canaan, honey flows less swiftly than milk. Yet with these polymers one observes precisely the opposite effect.

A colleague of mine, John Lumley, of Cornell University in the United States, decided to tackle this paradox.

A fluid flowing through a pipe moderately fast experiences eddies that can stir the fluid vigorously. But near the walls of the pipe there is a very peculiar region, called the turbulent boundary layer, where the eddies are "killed" by the wall—the fluid that is in direct contact with the pipe does not move, and the boundary layer is the transition zone into the turbulent flow. This zone plays a crucial role in determining the friction between the fluid and the wall. Lumley convinced himself that the presence of polymers in solution could alter the structure of this layer and, thereby, decrease the losses incurred by the flow as a whole. For a long time, I gave my students this explanation, which I found convincing. But, little by little, doubts crept in. I began to consider a different hypothesis. Could the small clumps of polyox, jostled and stretched by the current, act as miniature springs in the fluid, causing an elastic effect similar to that in the tubeless siphon?

Nothing could resolve this Byzantine puzzle. Interestingly enough, progress came not from theories, but from a clever experiment conducted in Germany. The experiment involved a turbulent flow created in a fairly long pipe (several meters in length) (Figure 7). In order to quantify the friction experienced by the flow, the pressure in the fluid

Figure 7

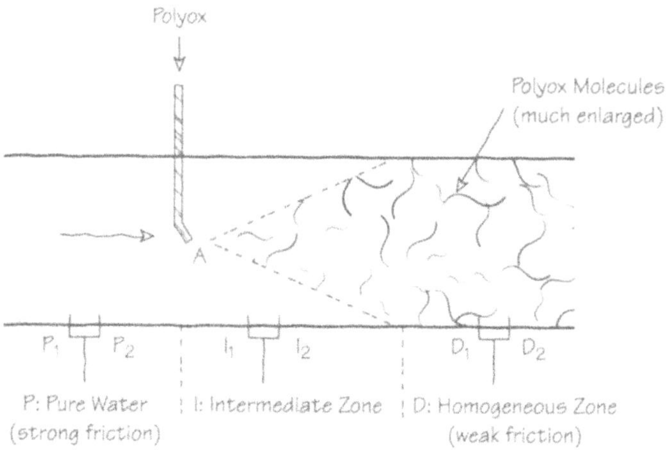

A crucial experiment to resolve a debate.

can be monitored at two neighboring points, for example at P_1 and P_2. The pressure difference between the two points provides a measure of the friction. A polymer is injected through a nozzle, connected to a small tube, at point *A*, located at the center of the pipe. And here is the subtlety of the experiment: The long chains which exit the nozzle are carried by the fluid. Churned up by the turbulence, they gradually fill, after a distance of a few meters, the entire cross section of the pipe. The flow is fast enough for the polyox molecules to be prevented from moving back against the current.

Three distinct flow zones can be identified.

- A zone of pure water (*P*) upstream of the nozzle. By measuring the pressure difference between P_1 and P_2, one finds that the local value of the friction is high.
- A zone (*D*) far downstream of the nozzle, where the polymer is fairly homogeneously mixed with the water (due to the turbulence of the flow). Here the measured friction is weak. This is in agreement with the various observations described so far (i.e., firemen, sewers).

- Finally, the last zone (*I*), intermediate between the injection point *A* of the polymer and zone *D*. Here, the polymer molecules have begun to fill the space in the tube, but have not yet reached all the way to the walls. Nevertheless, one observes (with surprise) that the drag has already been reduced substantially, in spite of the fact that in this intermediate zone the polymer has not had a chance to reach the turbulent boundary layer.

The conclusion is inescapable: the reduction in drag cannot be a property of the boundary layer.

Questions and Answers

I like to describe to high school students this experiment on drag reduction. The experiment itself has resulted in more progress on the issue than 10 years of theoretical debate on a phenomenon known for almost 50 years, which still remains poorly understood.

My purpose in using this example, is to show that our educational philosophy in France, dominated as it is by theory, is prone to distorting reality. The numerous questions asked by students and teachers alike show that this debate—theory versus experiment—is a central one. I will return to it in later chapters devoted to research and the education system. But right here I want to underline a few points.

Four Equations of Diamondlike Purity

In France, education is largely dominated by theory. In contrast to the style in Anglo-Saxon countries, the teaching of the sciences here is rather dogmatic. It makes students believe that knowing a few theorems, a few general principles, is enough to explain every phenomenon, and even to invent a few applications—which, by the way, has never been the primary preoccupation of our educational system.

Consider, for example, Maxwell's equations. Around 1867, the English physicist James Clerk Maxwell (1831–79) summarized in four lines of equations every phenomenon known at the time involving currents, electric fields, and magnetic fields. These four equations are nothing short of a pure jewel! At least in the way they are formulated today: elegant and concise. The temptation (to which we are all vulnerable) is to present them in their most complete form, and to deduce from them every single property which they contain: the action of magnet on a wire carrying an electric current; the propagation of radio waves or of light—electromagnetic waves—not to mention the dynamo, the electromagnetic braking system of trucks, even the microwave oven!

But what is rarely shared with the student is that these equations did not fall from the sky, that Maxwell had to go through a tortuous mental process and overcome many obstacles, before he could work out these equations based on observed phenomena. He started by writing them in approximate form; he groped, corrected himself, added terms to make the formulas more coherent. Only then did he find that these terms fit in a larger framework, which made it possible to predict new phenomena. Maxwell then realized that these equations, derived on the basis of phenomena that are constant or slowly varying in time, describe equally well the transmission through space of rapidly oscillating electric and magnetic fields. Starting from electrical and magnetic quantities measured in the laboratory, he was then in a position to calculate their speed of propagation. The result of the calculation was approximately 300,000 kilometers per second in vacuum (or in air). At about the same time, the French physicists L. Fizeau (1819–96) and J.-B. Foucault (1819–68) experimentally measured the speed of light and came up with a very similar value. Eureka! Maxwell concluded that light had to be one of those waves hidden within these equations. Shortly thereafter, the German physicist H.R. Hertz (1857–94) confirmed Maxwell's conclusion by demonstrating the existence of waves that are clearly electromagnetic in origin, but in other ways are similar to light: radio waves.

The True Birth of Ideas

Students faced with Maxwell's equations are at risk of being intimidated, indeed inhibited, by the perfection of such a construct. They might tend to think that they will never be able to match that accomplishment, and are likely to shy away from scientific research. It is our job to reassure them and boost their morale by sharing with them the vagaries of our profession. To that end, it is relevant to delve a bit into the history of the sciences.

A secondary education excessively biased toward the deductive discourages many students and stifles their inquisitiveness and their thirst for knowledge. Not every one of them proves capable of reasoning right away about abstract entities, which causes in many a mental block vis à vis the sciences. The question they most often ask is: What good is it?

It is obvious that a scientific education is not merely a voluminous catalog, with subdivisions entitled "electricity," "mechanics," "plastics," and so forth. Given the large volume of knowledge to be taught, one cannot forego the great, all-encompassing equations. Besides, they are so esthetically pleasing. But I do believe it essential to show, at the same time, that they are not the "tablets of the ten commandments." I do believe it is very important to emphasize to young people that the development of new ideas is not a linear path, but that it is filled with detours and obstacles. It is imperative to acknowledge that even the great minds, the Maxwells, the Newtons, had to go through many trials and errors while building their theoretical edifices. Even they could make mistakes. After the brilliant period of his discoveries in optics, gravitation, and mathematics, Newton himself became side-tracked for the next 10 years in alchemy.

By way of conclusion, I would like to quote the last sentence from a lecture by Max Born (1882–1970) entitled "Experiment and Theory in Physics." Born, one of the great theoreticians of modern times, gave this lecture in England in 1943. He said:

> To those who want to learn the art of scientific prediction, my advice is not to confine themselves to abstract reasoning, but to endeavor to decipher Nature's secret language as it is conveyed in natural documents: experimental facts.

CHAPTER 4

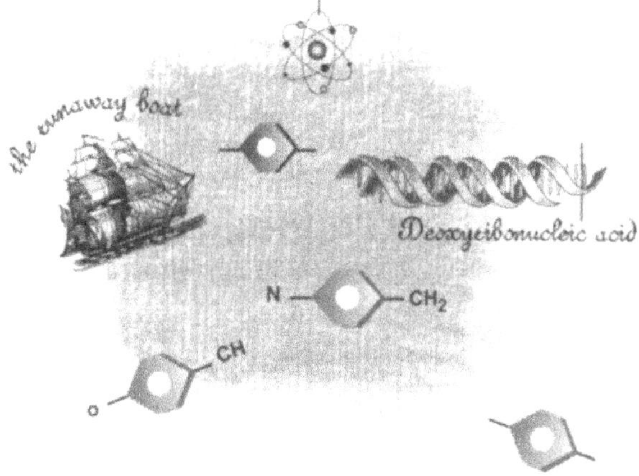

The Egyptian Scribe, Arabic Gum, and Chinese Ink

Chinese ink, what the British call *India ink*, actually came to us from Egypt. It was a remarkable discovery, prompted some 4000 years ago by the desire to write.

Carbon Black

It is likely that, from the time of the first cave paintings, people noticed that in order to leave a record on a surface, in order to draw, it was convenient to use a colored liquid. To make what today we call ink or paint, the simplest technique, indeed the most obvious one, is to disperse in water a natural, finely pulverized color pigment. Carbon black or charcoal for blacks and natural oxides for browns, yellows, and reds fill that need quite well. Using a liquid offers several advantages. With a wooden stick or a cut reed, with a bird quill or, later on, a metallic pen, with a fur pad or a hair brush, it is easy to deposit the ink or the paint on a porous substrate, like wood, stone, papyrus, or paper. The liquid serves another useful purpose. It wets the surface and soaks the substrate, dragging along with it the colored grains, which get encrusted in the solid. The finer the size of the grains, the more permanent the penetration into the substrate, which ensures the durability of the coloring after the water has evaporated: the colored grains will not be blown away at the first gust of wind, as they would had they not impregnated the underlying substrate. But the technique has a serious drawback: the ink must be used shortly after its preparation. It is not stable.

Picture a scribe from the ancient Egyptian empire preparing black ink with this process. He first lets the flame of a candle brush against a solid block, and he obtains a carbon deposit of extremely fine particles: carbon black (Figure 8*a*). He then scrapes those black particles into water (Figure 8*b*), and shakes the liquid vigorously. The result is black ink. One unfortunate day, after dispersing the carbon black in water, our scribe finds himself with a colorless liquid and, at the bottom of the

Figure 8

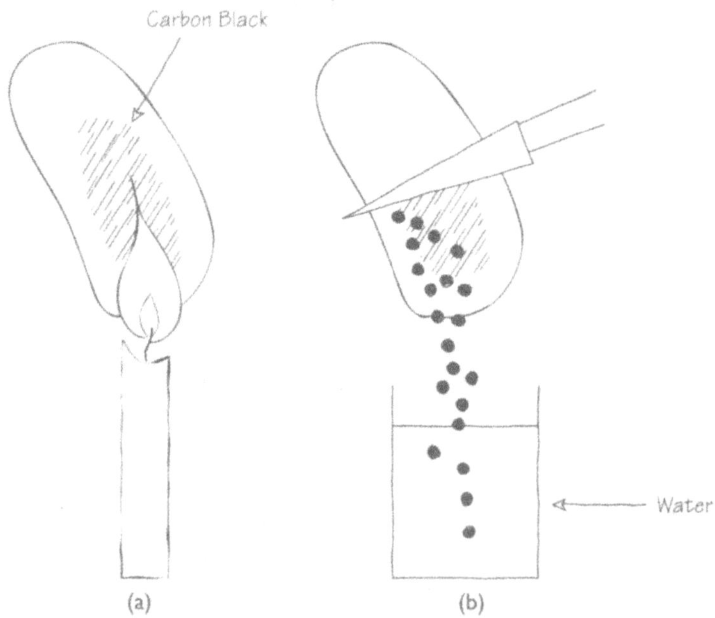

A primitive recipe for black ink.

Figure 9

Nature does not always cooperate.

container, a black deposit very difficult to disperse (Figure 9). The entire operation must be redone.

Who was the brilliant scribe of the second millennium who was inspired to dissolve in water a pinch of Arabic gum, and then to disperse the carbon black in that solution? Nobody knows, but the result is remarkable: the ink remains stable for at least a year (Figure 10). This process is still used nowadays in the preparation of watercolors.

Impossible Loves

Two questions come to mind:

- Why do the small grains of carbon black clump together to form large grains that fall to the bottom of the ink container?

Figure 10

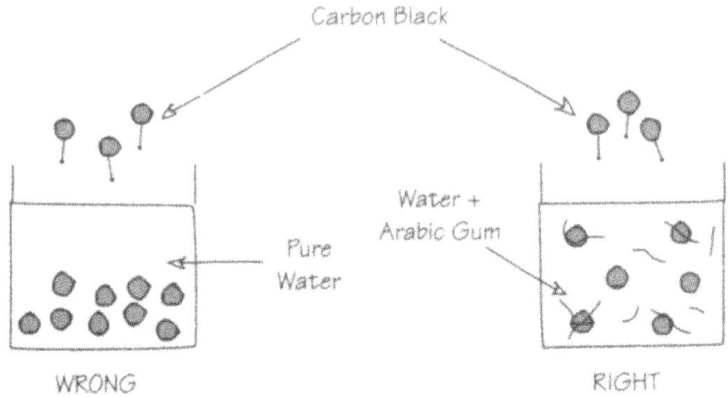

An improved recipe for black ink.

- Why does Arabic gum interfere with the *flocculation* process, as this aggregation into flakes is called?

To answer those questions, let us start by dissolving a little alcohol in water. All the molecules, those of water as well as of alcohol, move about in all directions, collide incessantly with one another like grains of wheat shaken in a sieve. This is due to thermal agitation, which promotes mixing. But two alcohol molecules also experience the action of another force which draws them toward one another, the so-called van der Waals force, first proposed at the beginning of the century by the Dutch physicist of that name. In the absence of thermal agitation, the two alcohol molecules would be attracted to each other under the influence of van der Waals forces, and would end up getting stuck to each other. But these two molecules are also subjected to many collisions with the water molecules surrounding them, and this thermal agitation keeps them apart.

But what about our grains? They are very much larger than an alcohol molecule (from 5000 to 10,000 times larger), and that changes everything. The van der Waals attraction overwhelmingly outweighs the thermal agitation. Thus when two grains are bound together, the energy is large, some 100 times larger than the binding energy of two molecules in contact. The collisions due to thermal motion are no longer energetic enough to separate them. One might think that the cumulative action of 100 collisions could succeed in separating the two grains. But collisions take place in random directions, and they tend to cancel each other out.

Eventually, the grains attract other grains to form flakes, enormous and fragile objects interconnected much like a corral. Their sheer weight causes them to drop to the bottom. So why does Arabic gum impede this flocculation?

Arabic gum is extracted from the sap of the acacia tree, a thorny species quite prevalent in semiarid regions. The gum contains long-chained sugar molecules: polyhyaluronic acid. These molecules, readily soluble in water, also stick easily to the grains of carbon black. But since they compete with one another, a single molecule cannot completely surround a grain, but sticks to it at only a few points. Each grain can thus bond to many molecules creating something resembling a forest of hairs, called a corona, on the surface of a grain (Figure 11). The name comes from the appearance of this crown-like structure. (In cross-section in a

Figure 11

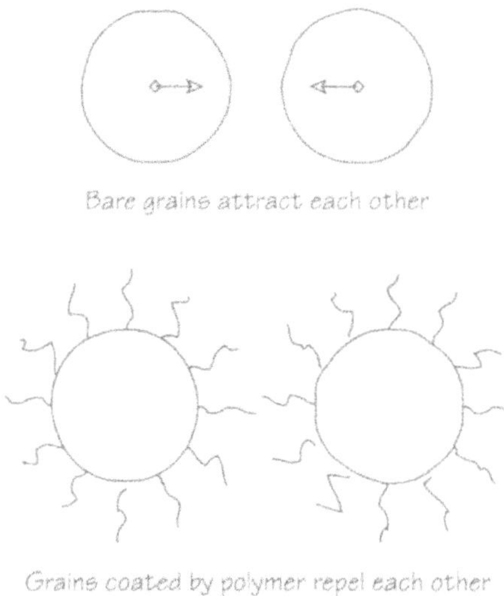

Stabilization of a colloidal suspension via polymer coronas offsetting the van der Waals attractive forces.

microscope it is reminiscent of the solar corona visible during a total solar eclipse, when the moon completely obscures the solar disk.)

This is not to say that the van der Waals forces have disappeared. They do continue to act, and two grains attract each other as strongly as ever. However, when they approach each other, it is their two coronas, with their sugar macromolecules, which come into contact. But since these macromolecules are soluble in water—"hydrophilic," in the terminology of scientists—their bonds to the water molecules are stronger than their van der Waals attractions to each other. We could say they "like" being surrounded by water and "prefer" that to coming into contact with other macromolecules. The resulting forces of repulsion between the coronas prevent them from getting too close to one another. That is how grains of carbon black remain isolated. The colloidal suspension (an ultra finely dispersed material) of carbon black is said to have been *stabilized.*

The Additive That Makes a Difference

Four thousand years separate the invention of Chinese ink and a complete explanation, which is only about 10 years old, of the mechanism of stabilization provided by the polymer corona. Here again, the invention preceded—by a very long stretch indeed—the explanation! And in this case too, the explanation was the result of a long, tortuous path, marked by successive trials and errors as well as by discoveries.

At this point, I must describe a remarkable experiment done by the British chemist and physicist Michael Faraday (1791–1867), one of our "great ancestors." We owe him the mechanism of electrolysis, the first liquefaction of numerous gases, and the laws of electrical induction.

So here is Faraday preparing a colloidal suspension of gold. He rubs under water two gold electrodes connected to a battery. The resulting sparks extract extremely fine particles of gold which remain in suspension in water and give the liquid a lovely reddish color. As it turns out, this colloid is stable. Even though the grains are subjected to van der Waals attraction, they remain separated.

Through this experiment, Faraday has just discovered an alternative mechanism for colloidal stabilization: electrostatic interaction. Gold

Figure 12

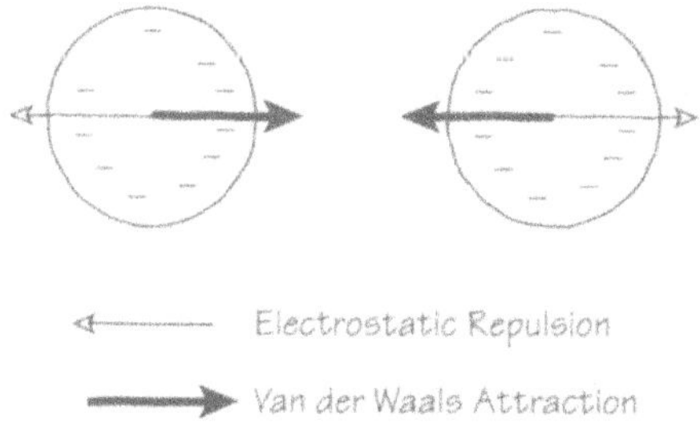

Electrostatic stabilization of a colloidal suspension.

particles happen to spontaneously carry negative electrical charges (Figure 12). It is well known that two electrical charges of like sign repel each other. This electrostatic force neutralizes the van der Waals attraction, and the grains remain isolated: the colloid is stabilized.

To demonstrate this, Faraday dissolves salt (sodium chloride) in the suspension. The color changes to blue: the grains have aggregated into larger structures. What has happened? Sodium chloride molecules (NaCl) dissolved in water dissociate into their two constituent atoms—one is positive (Na^+), the other negative (Cl^-). The positive sodium ions (Na^+) approach the negative grains (two opposite charges attract each other) and neutralize the electric field of each grain. Only the van der Waals forces remain (Figure 13), which cause the grains to attract each other and to clump together. The optical characteristics of the suspension are altered: the color turns blue.

Figure 13

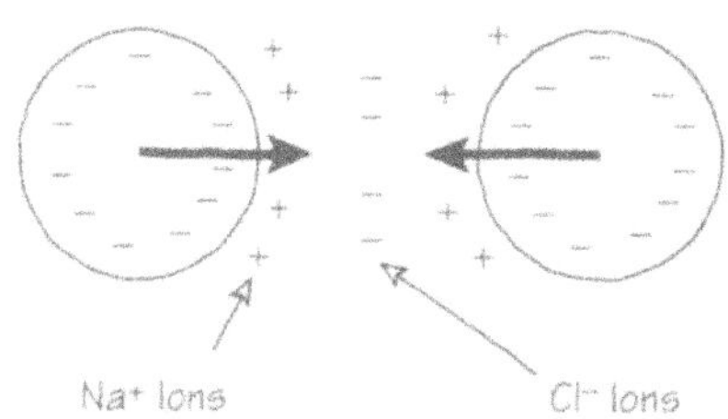

Inhibition of the stabilization process by dissolution of an ionic salt.

But Faraday performs yet another experiment. In the stable colloidal suspension of gold, red in color, he first dissolves a small amount of *gelatin*, a substance of long-chained molecules which are hydrophilic and which play a role similar to that of the sugars in Arabic gum. When he subsequently adds salt to the solution, the colloid now stays red; it remains stable. Faraday has just rediscovered the Egyptian method for stabilizing a colloid: the corona. The red colloid of gold to which he adds gelatin is doubly stabilized: once by the electric charges carried by the grain and once by the gelatinous corona surrounding each grain. While the salt added at this point suppresses the electrical interaction as it did before, the corona is still active, and the colloid remains stabilized, just as it does in Chinese ink.

I consider the story of Chinese ink a perfect example in more than one respect. It illustrates the properties of finely divided matter, which plays so prominent a role in our daily lives. Witness the products of the food industry (creams, margarine, mayonnaise), of the oil industry, of the cosmetic industry, and of many others. It also provides another example of the radical changes in physical properties which can be imparted by a seemingly weak action, in this case the addition of small quantities of polymers. This effect is the principal characteristic of what I have already called soft matter.

Paint, the Magic Potion

I also like the story of Chinese ink because it shows that even a product so trivial in appearance that we no longer accord it any attention has the power to amaze us by the marvelous quality of the invention which gave birth to it and by the subtlety of the physical phenomena which explain its behavior. We should be equally amazed by many of the industrial products which we use in our daily lives.

Take the example of white paint which we use to coat a wall or a ceiling. It contains tiny grains of titanium oxide (one of the most widely used white pigments). As you might expect, the suspension is stabilized with a polymer, which makes it viscous. But for the paint to be applied smoothly, the effort required of the painter as he presses his roller onto the wall must not be too great. Under the mechanical action of the roller against the wall, the paint should become fluid, not unlike a yogurt stirred with a spoon. Yet as soon as the roller has passed, the layer of paint should revert back to a viscous state, so it will not run. Here are three successive and contradictory requirements (among the total of five or six demanded of a modern paint). All of them can be satisfied with suitably chosen additives!

Questions and Answers

Rather than inquiries of a general nature, this account has prompted primarily scientific or technical questions. We highlight two of them, fairly representative: What is the difference between van der Waals forces and gravitational forces? Do the latter play a role between grains of matter? What is the difference between flocculation and coalescence?

- Gravitational forces are extremely weak. They vary slowly with distance and are proportional to the masses of the interacting objects. Therefore, they are the dominant factor for enormous objects separated by vast distances, such as planets and galaxies.
- Van der Waals forces decrease more rapidly with distance; they are important only for separations less than a micron.
- Flocculation is nothing more than the phenomenon that I have described in connection with Chinese ink. The grains of carbon black attract each other and, barring the action of a stabilizer, stick together. When the grains become large enough, they drop to the bottom of the container.
- Coalescence, on the other hand, refers to the union of droplets of liquids (not solid grains like those of carbon black or gold) in suspension in another liquid.

When two liquids are immiscible, as are water and oil, it is sometimes possible to suspend droplets of one in the other. One can disperse very fine water droplets, for instance, in oil (Figure 14). Such suspensions, also called *emulsions*, are prevalent around us: vigorously shaken vinaigrette sauce, mayonnaise, hollandaise sauce. In order to stabilize

Figure 14

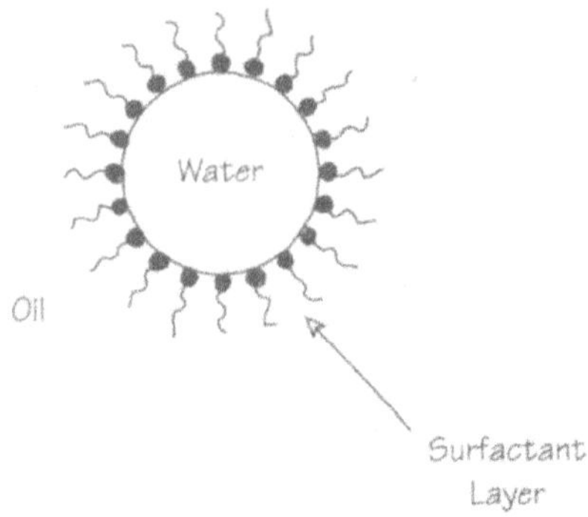

The action of a surfactant on an emulsion of water in oil.

such an emulsion, one uses a surfactant, which is composed of long-chained molecules, one end of which "likes" water and the other end of which likes oil. (A surfactant is the active ingredient in soaps and detergents.) The surfactant layer stabilizes the water droplet, whose outer layer is made up of hydrophilic groups. By contrast, oil molecules only "see" those ends of the surfactant molecules which resemble them. A surfactant molecule is only about 20 angstroms (Å) [4] long—which is approximately the thickness of the protective layer—and 5 Å in diameter. When two water drops approach each other, they coalesce by fusing into a larger drop. What makes this possible is that neither the drops nor the surfactant layers are rigid. Their shape gets deformed under the influence of the collisions associated with thermal agitation. Coalescence involves "holes" 30 Å in diameter, which corresponds to about one hundred molecules.

Although one is often concerned with stabilizing emulsions or suspensions, it is also sometimes necessary to do the opposite, that is, to destabilize them, to break them. For instance, in purifying water, it is useful to cause the particles in suspension to flocculate, and subsequently filter them out. To a suspension that is too stable, one can add a very small amount of a polymer to prevent the formation of complete coronas. A polymer molecule will sometimes stick to two or more particles, rather than to a single one. The "necklace" of particles formed in this way can rapidly evolve into a mass sufficient for flocculation to occur. This process might one day turn out to be useful for water treatment. The challenge is to find additive molecules that have no toxicity associated with either the additive itself or its degradation products.

Another possible application is in oil refining techniques. Crude oil is a viscous mixture of light and heavy products. The former yield fuels as well as compounds used in chemistry. The latter (which may even contain solid grains) provide tar and bituminous substances. In a refinery, the mixture is heated in what amounts to a large still to separate the different fractions. The lighter parts evaporate, are collected at the top of the distillation column, and are subsequently condensed. The heavy parts remain at the bottom, where they are recovered at the completion of the operation. This rather brute-force method is a voracious consumer of energy. It is conceivable that, in a few years, crude oil might first be treated with suitable polymers to bind together the solid particles, as in the purification of water, which could then be separated by gravity. This would make for an easy recovery of the asphaltenes, which are a primary ingredient of the material used in surfacing our roads. The remaining liquid, now much cleaner, could be treated easily to separate the light products themselves.

CHAPTER 5

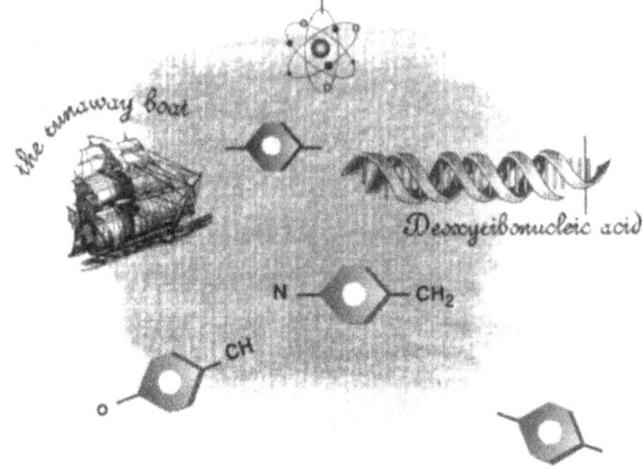

Liquid Crystals and the School of Fish

The Three States (of Matter)

We have all learned in school that matter presents itself in three different states: solid, liquid, and gas.

- *Solids* retain a permanent shape. Their atoms are close to one another and form a rigid lattice. They are often anisotropic, that is, their properties vary depending on the direction along which they are measured. For instance, in some solids the sound velocity depends on the propagation direction. In crystals, which are an "ideal" form of a solid, atoms are arranged in a regular pattern (Figure 15), and the anisotropy is a direct consequence of this ordered structure.
- *Liquids*, by contrast, have no shape of their own; they simply adopt the shape of their container. They are easily deformed by weak forces. They are *isotropic:* their properties are the same regardless of the measurement direction. The molecules in a

Figure 15

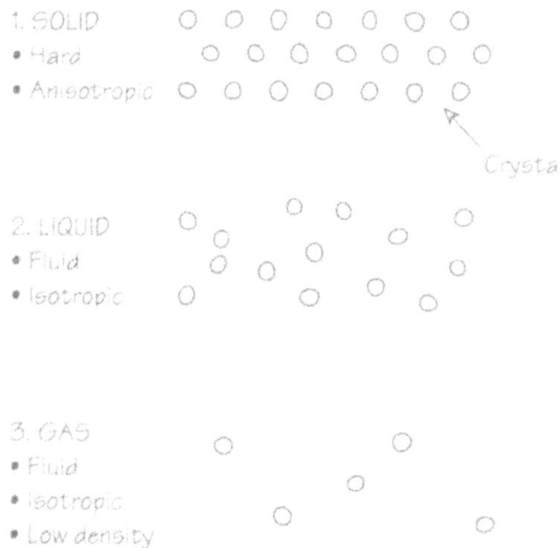

The three ordinary states of matter.

liquid are not fixed. Rather, they are in constant motion because of thermal agitation. Liquids are *disordered.*

- *Gases* are also fluids. They occupy whatever volume is available to them. Gases are also disordered, but their molecules are generally much farther apart than in a liquid. The density of a gas is typically a thousand times smaller than that of a liquid. The intermolecular forces are very weak. The only important forces are those which result from collisions between molecules.

But this classification scheme turns out to be too simplistic. There exist many intermediate states between liquids and solids: for example, liquid crystals. Discovered a century ago, liquid crystals have in the last 20 years become familiar objects. For example, they are the active part of the display of digital watches and pocket calculators. Let us stop a moment to consider this marvel: a tiny electric battery powers our watch for a full 2 years, while each second a new black digit, easily readable, shows up on a white or gray background. *A minute quantity of energy* is sufficient to cause the display to change.

Figure 16

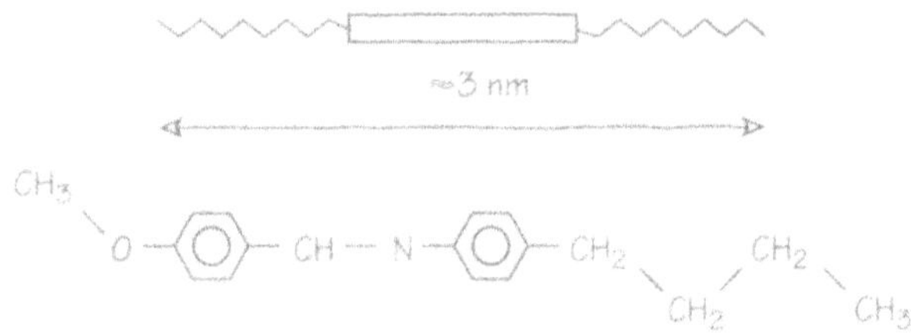

Structure of the molecule of a common liquid crystal, MBBA.

Judiciously Chosen Molecules

To obtain liquid crystals, one most often starts with elongated molecules. Figure 16 illustrates the currently accepted model. The rigid, central part would not alone form a liquid crystal. If our molecules were perfectly rigid, they would form a crystal rather than a liquid at ordinary temperatures. Hence the two flexible chains at either end, which favor the liquid state, are needed.

Figure 16 shows a liquid crystal of this type—its chemical name is N-(4-methoxybenzylidene)-4-butylaniline, but it is called MBBA in the jargon of the trade.

The Nematic State

In an isotropic liquid, the molecules, even when elongated, point in all directions of space. It is this very disorder which makes the fluid isotropic. In a nematic liquid (Figure 17), the elongated molecules remain parallel in a common orientation, much like a school of fish. But the relative positions of the different molecules are not fixed, and the molecules are free to move. One is dealing with a liquid, albeit an anisotropic one. The properties of a nematic liquid depend on whether they are measured in a direction parallel or perpendicular to the molecular axis (directions 1 and 2, respectively, in Figure 17). The direction of alignment itself is influenced by external conditions, for example, grooves etched on the bottom of the container.

Figure 17

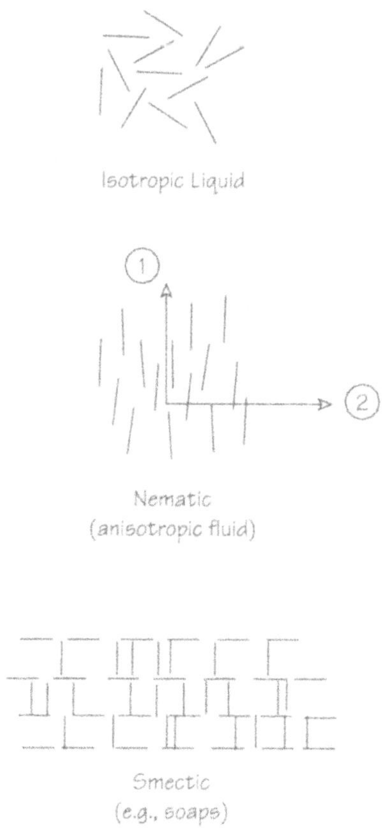

Differences between an ordinary liquid and two types of liquid crystals: nematics and smectics.

The Smectic State

Smectic liquids exhibit a very different structure. The molecules are arranged in layers (Figure 17), somewhat like a Napoleon pastry. Soaps have this structure: hence the term *smectic*, from the Greek *smêktikos* ("soap"). The thickness of each layer is well defined. Within a given layer, the molecules are reasonably parallel to each other and form a two-dimensional liquid: the molecules are confined to their assigned layer but retain the freedom to move within it under the influence of thermal agitation. The centers of the molecules have a disordered distribution.

In the main family of smectics, the molecules are perpendicular to the plane of the layers. In the other, less common, smectic crystals, the molecules lie in the plane of the layers. Like nematics, these liquid crystals are anisotropic: their properties are different when measured in a direction parallel or perpendicular to the layers.

How to Recognize a Liquid Crystal?

The first individual to really understand the nature of a liquid crystal was Georges Friedel (1865–1933), part of the second generation of an extraordinary lineage of scientists. In the nineteenth century, this family of Alsatian protestants, refugees of the war of 1870, first gave us a great chemist, Charles Friedel (1832–99). A fundamental reaction of organic chemistry bears his name: the Friedel-Kraft reaction. Another Friedel, from the third generation, became director of the School of Mines in Paris. And Jacques Friedel, representing the fourth generation, is currently making his mark as a pioneer of the science of metals in France.

In the 1920s, with the aid of an ordinary microscope, Georges Friedel succeeded in understanding the spatial arrangement of molecules that he could not see directly (they are only 20–40 Å long or about 2–4 millionths of a millimeter). From the observation of certain lines, only a few tenths of a millimeter long, Friedel deduced the structure of smectics.

More precisely, it was the defects in the molecular arrangement, as they appeared in the field of view of the microscope, that Georges Friedel studied. My earlier description of liquid crystals was admittedly oversimplified. What I said is correct over a distance of a few tenths or hundredths of a millimeter. On that scale, the molecules of a nematic do indeed remain parallel among themselves. But over larger distances, this privileged direction can change.

Consider the schematic diagram of a nematic shown in Figure 18. In a region labeled *R* in the figure of one sheet of molecules, the molecules remain approximately parallel to a common direction. This di-

Figure 18

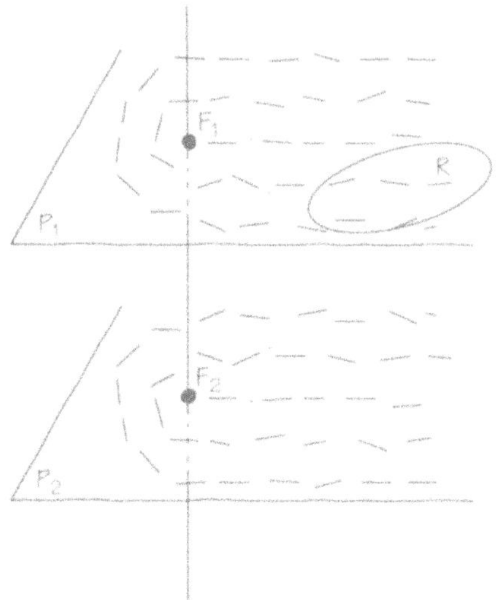

Three-dimensional configuration of a nematic.

rection eventually makes a turn as we travel some distance along with our school of molecular fish. Furthermore, the central line of fish, squeezed by the first curve, bumps against it (at the point labeled F_1 in the figure), not unlike codfish coming down the cold Labrador current and suddenly hitting the warm Gulf Stream.

The point where the molecules abruptly change directions over a short distance (as small as one molecular length), represents a sort of "catastrophe": it is called a singular point. A similar situation prevails in other planes parallel to this one, since molecules are parallel to each other not only in a particular plane but between two planes as well. The collection of singular points F_1, F_2 . . . forms a line, the appearance of which depends on the direction of observation: it looks like a wire when viewed from the side, but it looks like a point when viewed head on.

Smectics have a somewhat more complicated defect structure. If a layer of smectic liquid crystal is rolled tightly around to form a cylinder (Figure 19*a*), the axis of the cylinder forms a singular line *L*. Here the molecules abruptly change direction.

Figure 19

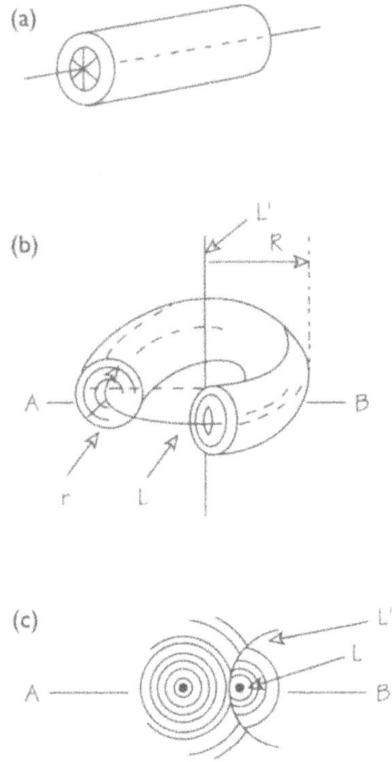

Two possible configurations of a smectic: *(a)* wrapped into a cylinder and *(b)* subsequently bent into a torus. *(c)* Pattern of singular lines associated with the torus, observed by Georges Friedel.

Now if the long axis of the cylinder is in turn bent to close on itself, one forms a torus, a geometrical figure reminiscent of a doughnut or a curtain ring (in Figure 19*b*, a section of this torus has been removed). The singular line has turned into a singular circle. Finally, if this ring is made thicker (Figure 19*b*) by stacking additional layers around the first one, the radius r of the toroidal cross section gradually increases. When it becomes equal to the radius R of the ring, the layers covering two diametrically opposed parts touch. At the point of contact exists a secondary singular line, which is the axis of the circle.

One can create more general types of defects by deforming the circle into an ellipse. The secondary singular line then becomes a hyperbola.

While observing these singular lines, Georges Friedel noticed ellipses and hyperbolas (Figure 19*c*). By studying the relative configuration of the ellipse and its associated hyperbola, he recognized them as the signature of the stacking of equidistant layers. It follows logically that a smectic is made of layers which can be bent, and this implies that the layers are liquid. A marvelous piece of deductive work!

Given the scant importance attached today to the teaching of geometry, would we be able to identify these singular curves, to recognize their significance, and to deduce from them the structure of smectics?

How to Issue Commands to a Liquid Crystal?

The primary application of liquid crystals is in the technology of displays (watches, electronic devices, etc.). The principle involved relies on the following properties.

- *Structural properties.* We have seen that the molecules of a nematic behave a bit like a school of fish, swimming parallel to one another. The direction of the school, which can be arbitrary, is often determined by some external agent, like grooves at the bottom of the container. Consider two parallel glass slides (Figure 20*a*), each lined with a set of grooves (labeled G_1 and G_2 in the figure); the slides are oriented so that the two sets of grooves are perpendicular to each other. A nematic inserted between the two slides is going to experience some "discomfort." Near the upper slide, the molecules want to orient themselves along grooves in that slide while near the lower slide they prefer to line up along the lower slide's grooves, which are perpendicular to G_1. Although the molecules would "like" to remain parallel to each other, they have to adapt to the external conditions. They adopt a compromise which best approximates their natural preference: they take on a helical configuration lining up as in a circular staircase (Figure 20*a*). This arrangement, first observed by Charles Mauguin, is called a twisted nematic.

Figure 20

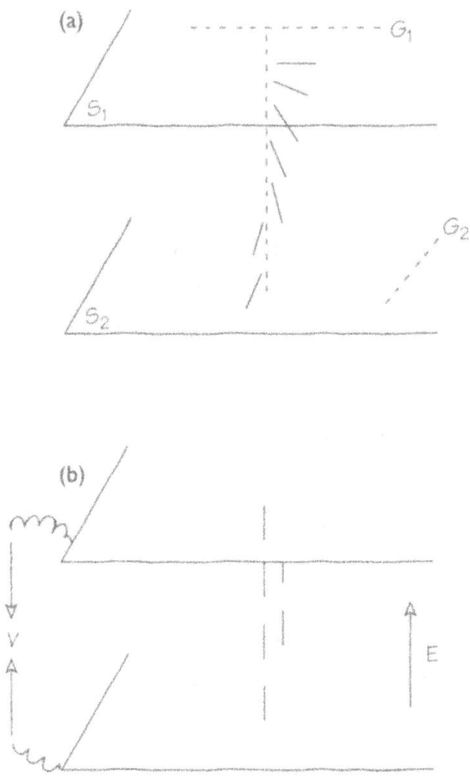

Spatial arrangement of a nematic layer contained between two grooved boundaries *(a)* at equilibrium and *(b)* after applying a perpendicular electric field.

- *Electrical properties.* When a voltage is applied between the two glass slides, an electric field is established.[1] The axis of the molecules then lines up parallel to the field (Figure 20b).
- *Optical properties.* The molecules have different properties parallel and perpendicular to their length. They are optically anisotropic. Polarized light enables us to take advantage of this anisotropy.

[1] To make the glass conductive, it is coated with tin oxide, a transparent material which conducts electricity.

If one examines the properties of a ray of sunlight, no privileged direction is apparent in a plane perpendicular to the ray (Figure 21*a*). The same is true of light emitted by ordinary sources, such as a candle or a light bulb. Things change, however, when sunlight is reflected by a glass surface. A reflected ray does have a privileged direction: it turns out to be the direction perpendicular to the plane formed by the incident and the reflected rays (Figure 21*b*). Light is then said to be ***polarized.*** It can be shown that the direction of polarization coincides with the direction of the electric field associated with the light wave.

Polarized light can be produced very conveniently by means of polarizers, the most common type of which comes in the form of a plastic sheet that is treated so that it allows through only light whose polarization direction is compatible with the characteristic direction of

Figure 21

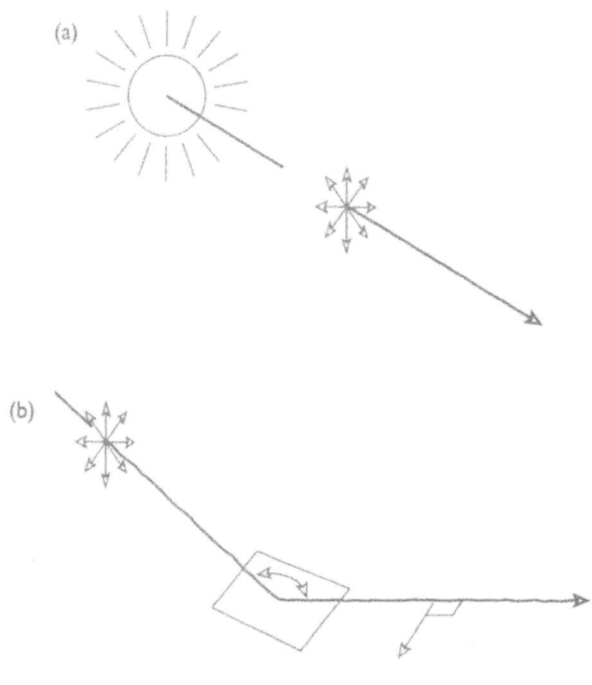

An ordinary light ray is unpolarized *(a)*. It can acquire a preferred polarization direction, for instance, by reflection off a planar surface *(b)*.

Figure 22

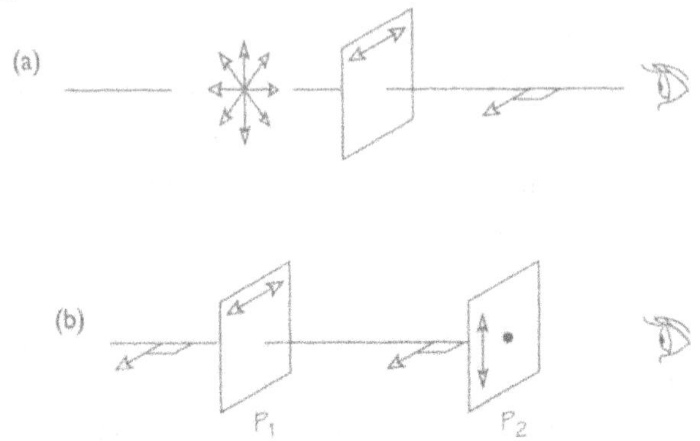

Manipulation of light with sheet polarizers. The light polarized by a first polarizer *(a)* can be blocked by a second one *(b)* whose characteristic axis is rotated by 90°.

the polarizing medium (Figure 22*a*). But a polarizer can also function as a filter. It either passes or blocks incoming light depending on whether the polarization direction is parallel or perpendicular to its characteristic direction (Figure 22*b*).

We can now explain how an LCD works. Two electrodes made of electrically conducting and suitably grooved glass (recall Figure 20) create a twisted nematic. A polarizer (labeled P_1 in Figure 23) produces polarized light, whose polarization direction is parallel to the filter direction of the polarizer. But the light is affected as it propagates through the nematic (Figure 23*a*). Its polarization direction twists around the ray's axis, *just like our body turns as we walk down a spiral staircase.* As light arrives at slide S_2, its polarization direction has become perpendicular to the filter direction of polarizer P_2 located just outside slide S_2. This second polarizer blocks the light, and the background appears black. If a voltage is now applied between the two electrodes, the resulting electric field causes the axis of the molecules to align along the direction of the field (Figure 20b).

Like an alpine mountaineer climbing along a rope, always facing the same direction, the light polarized by P_1 (Figure 23*b*) proceeds unaf-

Figure 23

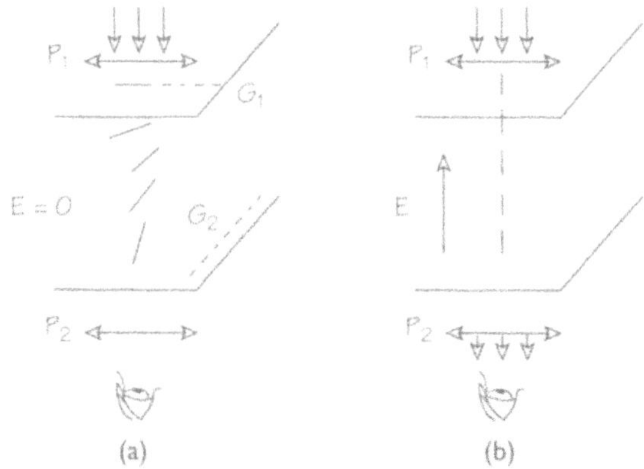

Implementation of a liquid crystal display with a twisted nematic between two polarizers. The background is dark at rest *(a)* and bright when an electric field is applied *(b)*.

fected through the nematic. The light eventually arrives at polarizer P_2 with a polarization direction that is unchanged. This direction is now parallel to the filter direction of the second polarizer P_2. The light passes through unimpeded, and the background remains bright.

In an actual display, the characters which are to show up (digits or letters) are created on glass slides. Each character is made up of short sections of straight lines in the form of a matching pair of electrodes in registration on each slide. Whether a voltage is applied or not between this pair of electrodes determines if the corresponding straight-line segment shows up or not.

Twisted nematic displays were invented by one of the great scientists of our time, W. Helfrich. These devices consume very little energy. Their principal limitation is speed: It takes a few thousands of a second after establishing or removing the electric field for the nematic molecules to switch direction. While this time is short enough for a watch or an electronic instrument display, it is far too long for a television screen. There is, however, some hope that other liquid crystals, more complex than the ones I have just described, might be suitable for that application: stay tuned.

CHAPTER 6

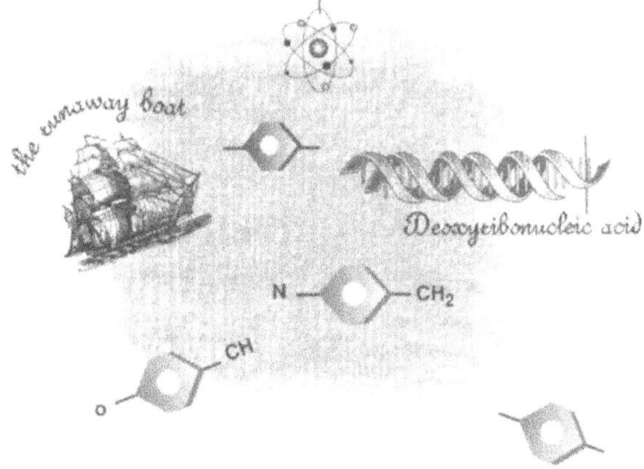

On the Surface of Things: Wetting and Dewetting

On a Duck Feather

The properties of surfaces play many practical roles: They are involved, for example, in problems of lubrication: the drop of oil that prevents a door from squeaking or permits a motor to turn at high revolutions and temperatures. It is also by means of surface effects that low-yield ores can be enriched. Surfaces mediate a variety of industrial chemical reactions.

A surface separates two media: a solid and a liquid, a liquid and a gas, etc. Let us first get acquainted with the affinity between a liquid and a solid by placing them in contact with each other.

We can take a drop of liquid (water) and deposit it on the surface of a solid. Depending on the choice of surface, we encounter two distinct situations.

1. The surface repels water; it is said to be *hydrophobic* (literally, "water-fearing"). The drop cannot spread, and instead assumes

the shape of a small, spherical cap.[1] This case is called *partial wetting* (Figure 24). It is observed, for instance, when a water drop is placed on a waxed surface or a duck feather.

2. The surface attracts water; it is hydrophilic ("water-loving"). This case is exemplified by extremely clean glass (or metal). The drop spreads, and ends up occupying a huge area by reducing its thickness down to extremely small values (comparable to the size of a single molecule). This situation is referred to as total wetting (Figure 24). Reaching the final state may require days or even weeks. During this time, some precautions must obviously be taken to prevent the water (or whatever liquid is used) from evaporating or becoming contaminated.

The science of wetting goes back to the very beginning of the nineteenth century (around 1805). We owe it to an Englishman, Thomas Young (1773–1829), and to his French rival, Pierre Simon de Laplace (1749–1827). Laplace was one of the most brilliant scientists of his

Figure 24

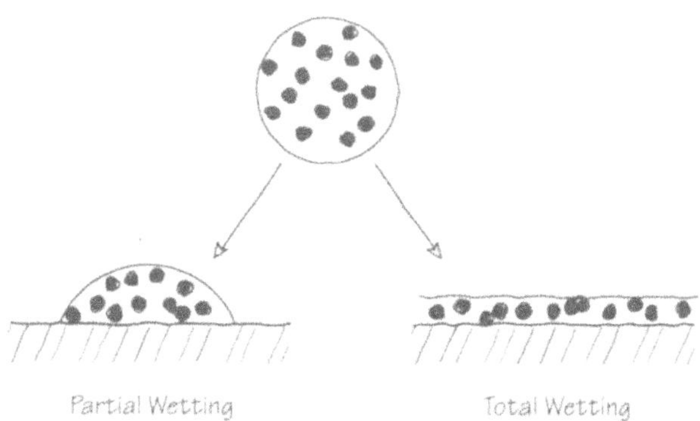

Partial and total wetting of a solid surface by a liquid drop (Thomas Young, circa 1805).

[1] This assumes that the drop is small (less than a millimeter), so as not to sag under its own weight.

generation: mathematician, physicist, and astronomer all in one. What distinguished them from each other was that Laplace relied on the modern mathematical methods of his time (partial differential equations), while Young was a more traditional practitioner of geometry, in the Greek tradition. For this reason, Laplace was often considered more "hip" by his contemporaries. It seems to me that they deserve equal credit.

In any event, the basic principles of the equilibrium state of liquid drops were worked out very early on. The relevant phenomena are controlled by the energy associated with the interfaces. This energy is related to the mutual interactions between the solid, liquid, and gaseous molecules that coexist in the vicinity of the interface. Young demonstrated that these interfacial energies determine the angle between the drop and the solid surface in the case of partial wetting.

Thomas Young started as a physician. He became interested in physics, and was to become one of the great physicists of his time, contributing to three different areas of science: wetting, elasticity (the science of the deformation of solids; there exists in this field a very important coefficient, called *Young's modulus*), and wave optics (which deals with interference phenomena of light: the experiment of Young's pinholes, demonstrated in every high school, constituted a defining moment in the history of the sciences; it involves light illuminating two small holes or slits in an opaque screen, which generates a characteristic interference pattern visible on a second screen).

Young's activities were not confined to physics. He also became involved in archeology. He studied Egyptian writings, finding himself once again in competition, this time with another Frenchman, J.-F. Champollion (1790–1832), who beat him by only a few months in deciphering hieroglyphs. Mr. Young, capable of simultaneously conducting research in such different disciplines, deserves our admiration. Could anybody be as eclectic nowadays?

A few years ago, our group at the Institute of Physics and Chemistry became very interested in the phenomenon of wetting. We wanted to study not only the final shape of the drop, but also how it gets there, that is, the dynamics of the spreading liquid. We were also seeking to establish the laws of the reverse phenomenon: *dewetting*. If one forces water to spread on a hydrophobic surface, how does the surface "dry" itself out?

I will illustrate the phenomenon of dewetting by describing a simple experiment. When you deposit a drop of water on a sheet of transparent plastic like polyethylene, you will observe that the drop does not spread. It is a classic case of partial wetting (Figure 25*a*). The polyethylene is hydrophobic. Imagine now that you mash the drop down with your finger, forcing it to spread on the plastic (Figure 25*b*). What happens then?

Figure 25

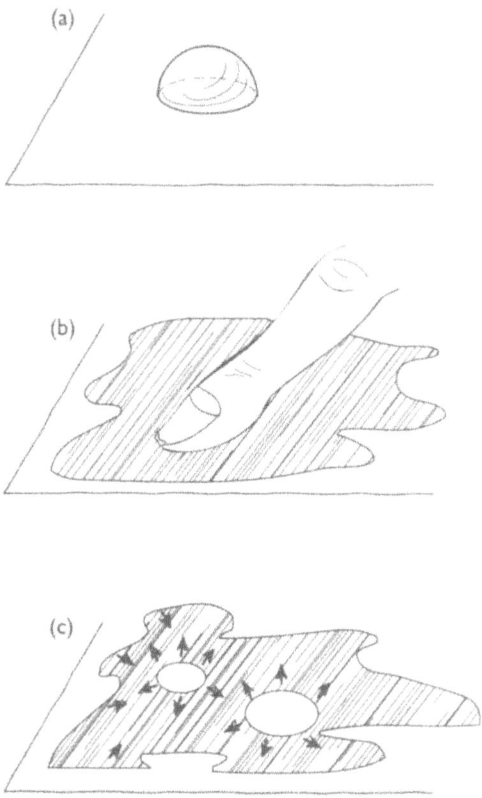

A dewetting experiment. Partial wetting causes a liquid to form a spherical drop *(a)*. If forced to spread on the solid surface *(b)*, the liquid attempts to retreat *(c)*.

Dry regions begin to form on the area covered with water. The dry holes expand as though the liquid were "fleeing" the solid (Figure 25*c*). Experiments demonstrate that the speed of expansion of the dry regions is well defined. The laws of dewetting were worked out quantitatively by F. Brochard, J.B. Brozka, and C. Redon. But they were not worked out without a struggle.

THE FAMILY RUG

Many phenomena, ranging from what happens on a road as rain begins to fall, to processes in chemical transformations, depend on the mechanics of dewetting.

One might wonder why these studies on wetting and dewetting did not follow closely on the heels of Laplace's and Young's discoveries. The reason is essentially one of technology. If the experiment is conducted under the crude conditions that I have described, the boundary between dry and wet regions turns out to be quite irregular. As the liquid flees the area on which you are pressing, the boundary seems to get stuck occasionally: It remains pinned at innumerable defects in the surface of the plastic. These could be physical imperfections (for example, scratches on the polyethylene) or chemical defects (like greasy spots left behind by a finger). In order to do the experiments reproducibly and to reveal the underlying laws, extremely clean and controlled surfaces are essential. In Young's time, such surfaces simply did not exist.

Today, we have at our disposal an extraordinary material called a silicon *wafer.* These very thin slices of silicon, roughly 10 centimeters in diameter, costing about $100 apiece, are the foundation of modern electronic microchip technology. A single crystal of silicon of that size features one hundred million rows of atoms, in precise and perfect order, one after the other, devoid of any flaw! This is a magnificent technological accomplishment. Having grown semiconductor crystals some years ago, while I was a student of Pierre Aigrain's, I can testify that such a result is not trivial to achieve.

One might think that such a beautiful object would also provide a flawless surface with which to conduct convincing dewetting experiments. Such is not, unfortunately, the case. This magnificent wafer is still marred by a few imperceptible surface defects: it is oxidized by air and is covered by a somewhat irregular film of silica.

What to do next? One can proceed the way a homeowner about to sell his house often does. When the wooden floor of the living room becomes damaged, an expedient solution is to buy a rug to cover the flaw. The same principle is used here: one tries to coat the surface of the wafer with molecules shaped like tiny hairs, enough molecules to cover the entire structure with a "rug" as uniform as possible, in order to hide the underlying defects. Still, it is not easy to create such a defect-free rug: water molecules tend to stick to the surface and strongly interfere with the deposition reaction. Here I cannot resist the pleasure of recalling the work of Jean Bruno Brozka, who has just completed his thesis at the Curie Institute on the topic of dewetting.

Brozka was working under difficult conditions. While going through school and preparing his thesis, he earned a living by working for the national meteorological office. He spent nights updating weather maps of the type shown on television. He usually showed up at the lab by morning's end, a bit tired and disheveled but motivated and inspired all the same. Brozka was told, "Your job is to produce wafers with nearly perfect surfaces." Nowadays, a young research scientist would be likely to reply, "But this reaction is extremely sensitive to contaminants, water, dust. . . . I am going to require a *clean room.*" (A "clean room" is a virtually dust-free space, where the atmosphere is controlled, which one enters only while wearing a mask; such a facility can easily cost $50,000.) Brozka thought about the problem, and tried several approaches over several years. But, in the end, instead of demanding a clean room, he suggested doing the experiment in the laboratory's courtyard, on sunny winter mornings. The purpose of that proposal was to take advantage of a condition well known to meteorologists: the lack of water vapor in the air. Using this simple strategy, and nothing more, he was able to prepare the most perfect layers known to the international scientific community at the time (1990).

I find this story instructive, not just because of Jean Bruno's personality, but also because it illustrates the kind of mind-set which is so needed in research: refrain from immediately resorting to a jackhammer to break open a hazelnut without first checking for an incipient crack on the surface of the shell.

What should also be appreciated in the story of dewetting are the relative contributions of theory and experiment. Understanding the theory of the growth of a dry hole is not easy, because each hole is surrounded by a liquid ridge, invisible to the naked eye, resulting from the accumulation of water extracted from the initially wet region. The structure of this ridge is rather complicated. Still, it took a team of two hydrodynamic theorists no more than six months to understand it, which is not all that much. By contrast, Brozka's dewetting experiments, performed with Claude Redon, represent four years of work. But the payoff was worth the effort: it uncovered the universal laws of dewetting.

A Pearl Necklace

In discussing the interaction between a solid and a liquid, I have described two possible wetting cases, with very different properties: partial or total. But is it always possible to tell one from the other? Experiments are generally unambiguous when a liquid is deposited on a relatively extended solid surface. Things are not so clear, however, when a thread is dipped in a liquid, a situation frequently encountered in the textile industry. Knowing the correct answer is of crucial importance.

To fabricate a textile thread from a melted polymer is quite an accomplishment. The viscous polymer is drawn at 6000 meters per minute through a nozzle with a small hole. It solidifies before being wound at high speed on a spool. For the thread to survive the spooling process, to slide smoothly on other threads during weaving, and to fix color pigments during dying, it is necessary to protect it by coating it with a liquid (this process is called oiling or greasing the thread).

The engineers in charge of solving this problem encountered a few surprises. The polymers used to make the fibers behave strangely. When they are in sheet form, the wetting is total, but threads made of the same material seem to be nonwettable! When dipped in a liquid, the threads emerge with a string of droplets, like the pearls of a necklace. One is reminded of the duck feathers. There seems to be a contradiction. Could it really be that this material ceases to be wettable when its shape changes from a sheet to a thread? Surely, the wetting characteristics should depend only on the physicochemical properties of the polymer and the liquid, and not on the geometrical configuration of the base material. Gradually, our theoreticians convinced themselves that the fiber really was wetted, completely coated with a liquid film in between the droplets. But how could they prove it?

The liquid film is invisible to the naked eye, and theoretical estimates of its thickness give a very small value—just a few nanometers. There are plenty of sophisticated and sensitive techniques for revealing the presence of such films. But I much prefer the experiment done by two alumni of the Institute of Physics and Chemistry: J.M. di Meglio and D. Queré. With simplicity and elegance, they were able to prove the existence of the film, and even to measure its thickness.

Cannibalistic Drops

The experiment involved taking a dry fiber, depositing two neighboring drops on its surface, about a millimeter apart (Figure 26*a*), and observing the result.

If the fiber is totally wettable, the liquid will start spreading from each drop and will end up covering the entire surface of the fiber, in particular the section between the two drops. It takes a while, but when the film is ready, a thin liquid cylinder actually connects the two drops. At this point, a rather amazing process begins: the smaller drop feeds the larger one (Figure 26*b*). So if we see an exchange of liquid taking place, there must be wetting, for how could a nonvolatile liquid transfer from one drop to the other without "following" the fiber, that is, without wetting it?

That the large drop devours the small one is, upon reflection, not so puzzling. We have all witnessed drops run along a window on a rainy

Figure 26

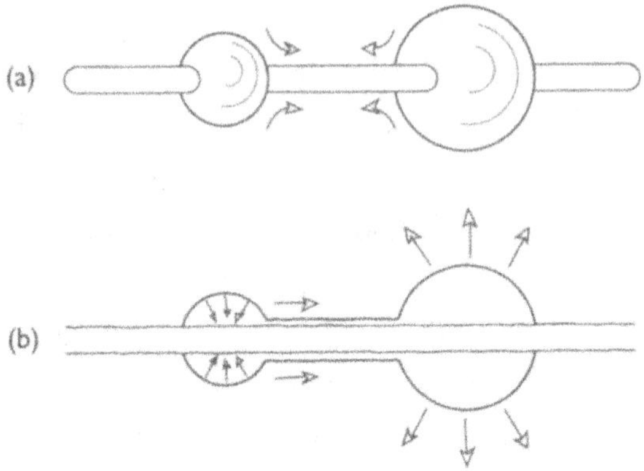

Transfer of liquid between two drops deposited on a solid thread.

day. We are familiar with the way two drops meet and fuse to form a larger one. It is easy to convince oneself that the total surface area ($s + s$) of two small drops is larger than that (S) of a single, larger drop entrapping the same volume of liquid (Figure 27). We will see later that there exists an energy proportional to the surface area of a liquid. Therefore, the most favorable configuration, that which involves the least surface energy, is one large drop, rather than two smaller ones. To reach that state, it is easier for the small one to feed its larger counterpart. It is this physical phenomenon of "cannibalism" between drops which guided our two experimentalists.

This simple experiment not only proved the existence of a wetting film, but also allowed its thickness to be determined. The time required

Figure 27

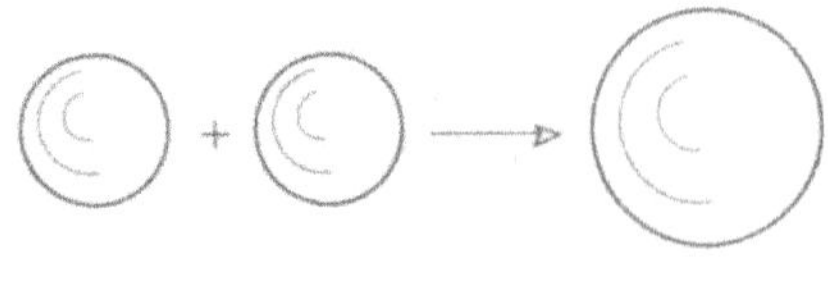

The surface of one large drop is smaller than that of two small ones of same total volume.

by the small drop to transfer to the large one is rather long: on the order of one afternoon, sometimes an entire day. The time depends primarily on the friction between the liquid and the fiber. The thinner the film, the greater the friction, and, in turn, the longer the transfer time; thus the transfer time provides a measure of the film's thickness.

The Benjamin Franklin Spirit

The two stories I have just told, of the silicon wafer and of the drops on a thread, highlight a frame of mind which is of great importance in scientific research. I call it the "Benjamin Franklin" spirit. To expand on what I mean by that, let us return to a surface—this time, the surface that separates a liquid from a gas. The interface is often the seat of bubbles and froth, which we will discuss in more detail later. For now, let us simply consider what happens to water when a small amount of a surfactant, such as soap, is added to it. Surfactant molecules are rather extraordinary objects.

They are relatively short (1 to 2 nanometers long), with violently antagonistic properties—almost schizophrenic, one might say. One end, often provided by an acid group, is strongly hydrophilic. We will call it the *polar head* of the molecule (Figure 28). The rest of the molecule is resolutely hydrophobic. It is a so-called *aliphatic chain*, made up of a string of 12 to 20 CH2 groups: effectively a small chunk of polyethylene [1].

Figure 28

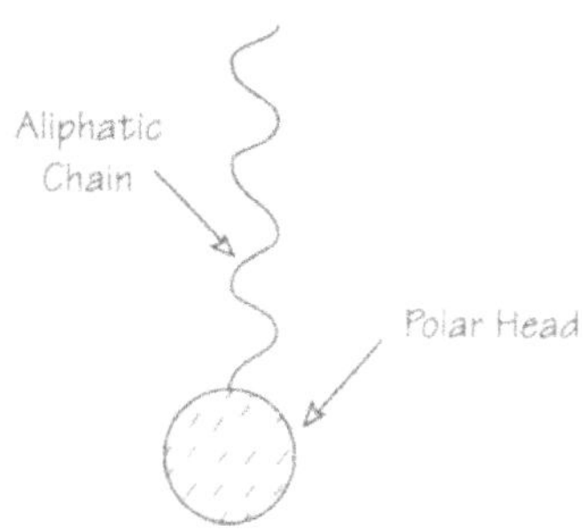

Structure of a molecule of surfactant.

If a single molecule of this type is immersed in water, it becomes extremely unhappy. Its aliphatic chain can think of only one thing: to escape the water that it so abhors. Helped by thermal agitation, it manages to reach the surface. There, the situation, while not ideal, is already much improved. The polar head can remain immersed with delight in the water, while the hydrophobic chain can dry itself out in the air. If there are many such molecules in the water, they can achieve an almost idyllic situation by squeezing against each other like penguins in a rookery: head in the water, and chain in the air, almost perpendicular to the surface. When the entire free surface of the water is covered, perfection is attained. The molecules form a very even layer (Figure 29), the thickness of which is equal to a molecular length. We have what is called a *monolayer.*

I sometimes ask my students: "How can one measure the thickness of a monolayer?" If they are nearing the end of their studies, they have become familiar with those marvelous tools at the disposal of modern scientists—X-ray generators, neutron reactors, and so forth—and after some thought, they might suggest using X-rays. And it is true that X-rays are rather well suited to this type of problem. But the best students will realize that the layer is quite thin. That means there is little material to interact with the beam and hence a very weak signal will be re-

Figure 29

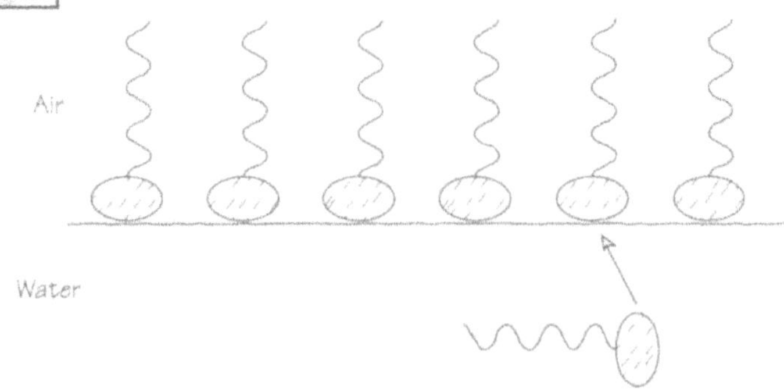

Optimum arrangement of surfactant molecules on the surface of water.

turned by the layer. The measurement would be possible with a synchrotron, which is a powerful X-ray generator. Of course, an experiment with this sort of hardware is extremely expensive.

Fortunately, 200 years ago, Benjamin Franklin devised a far simpler solution. We have all heard about Franklin, at least as the inventor of the lightning rod. He also was the ambassador to France from the young United States during the reign of Louis XVI. He was a man of great culture and great passion. He showed an interest in all manner of experiments done in the "Physics cabinets" in those days of the Age of Enlightenment. In particular, he was intrigued by the effects of oil on water. It had been known since the Greeks that a film of oil spread on the surface of the sea tends to quiet down the waves.

So, Franklin proceeded to the edge of a pond in Clapham, near London, and gently poured a spoonful of olive oil onto the water. Oil molecules are rather like those I have just described. The oil spreads, but the visual appearance of the surface does not change noticeably, because the layer of oil is very thin relative to the wavelength of light. Still, Franklin managed to recognize those areas which were covered: in the absence of oil, the breeze created wavelets on the surface of the pond, while in the presence of oil, the ripples were gone and the surface was smooth. It was as though the water's skin had become rigid! Based on this observation, Franklin was able to estimate fairly accurately the surface area of the oil film. It was enormous: on the order of 100 square meters.

Franklin's experiment carried the seeds of a profound result, which bore fruit, not for Franklin, but 100 years later for Lord Rayleigh.[2] If one divides the original volume of the oil by the surface area of the layer after it has spread, one finds the height of the film, which turns out to be of the order of a nanometer: this height is roughly *the size of the molecules of oil.* And it is this simple experiment—the beauty of which lies in the ability to observe and the insight to recognize the covered area—that our students should think of first.

---

[2] For a complete history of surfactant films, see Charles Tanford's admirable book *Ben Franklin Stilled the Waves* (Duke University Press, 1989).

This example of the "Franklin spirit" seems to me very important. The stories of the silicon wafers and of the liquid drops on the thread, which I told earlier, are also perfect examples of this Franklin spirit. Obviously, I am not claiming here that X-rays from the synchrotron, or that the neutron beam in reactors, are useless. Indeed, they are often indispensable. I myself have had to resort many times to neutrons to obtain the desired results! But the Franklin spirit should never be underestimated.

Surfactants have numerous practical applications. They have the ability to disperse in water substances which would otherwise be insoluble. That is the principle behind detergents.

A hydrophobic impurity gets hooked onto a fabric by molecular interactions (Figure 30*a*). Being hydrophobic, it is insoluble in water, and

Figure 30

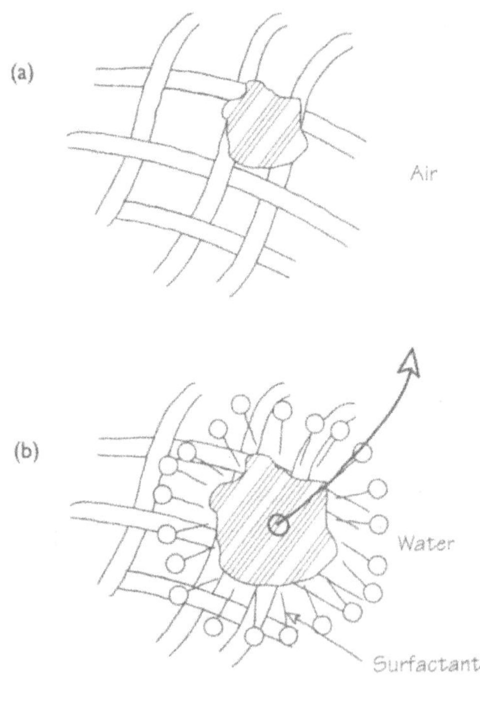

Action of a detergent. The foreign stain, bound to the fabric *(a)*, can move freely after being coated by a layer of surfactants *(b)*.

resists any attempt at washing in water. It is a familiar situation: a greasy stain on one's shirt after lunch. However, if a bit of surfactant is added to the water, its hydrophobic chains attach themselves to the impurity, while exposing toward the water a shell of polar heads which could not be happier than to swim freely away in the water (Figure 30*b*). And the stain is removed.

THE BILAYER AND THE RED BLOOD CELL

As mentioned earlier, the creation of a monolayer (a single molecular layer at the interface between water and air) results from the ambivalent nature of surfactant molecules. But surfactants have also devised another strategy toward happiness: they can form a bilayer (i.e., a double layer) within the water. They achieve this configuration by arranging themselves back to back. The aliphatic chains are quite comfortable being shielded from the water by their neighbors for whom they have an affinity. The polar heads form two hydrophilic surfaces, parallel to each other, blissfully immersed in water (Figure 31).

Figure 31

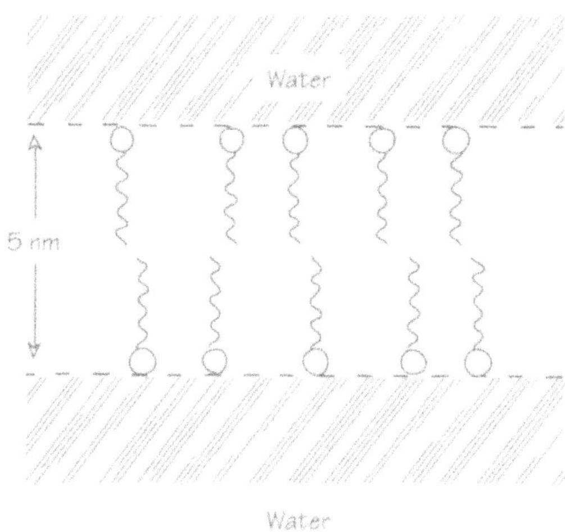

Structure of a bilayer.

Nature will perform wonders with these remarkable (but fragile) objects. The successive stacking of a bilayer, a layer of water, another bilayer, another layer of water, etc., forms a soap. We have already encountered a structure of this kind in our discussion of smectic liquid crystals. It is no accident that Georges Friedel gave them this name.

If such a double layer is folded back onto itself, one obtains a sort of a pouch. Figure 32 shows only a portion of the molecules, which actually cover the entire surface. The cells in our body are built in precisely this way. The walls enclose, of course, other molecules with specific functions, but their basic structure is essentially as just described.

Let us focus for a moment on a particular cell of this type: a red corpuscle in our blood. It is a small disk, biconcave in shape, about 20 microns in diameter, large enough to be visible in a microscope (Figure 33). It is a fairly special cell both in structure—it is rather degenerate, in that it lacks a nucleus—and in function: It plays a very important role transporting hemoglobin, a protein that fixes oxygen in our lungs and ferries it back to our tissues. This "object" is, essentially, an extremely deformable bilayer, which is a distinct advantage since it must be able to squeeze its way through the smallest blood vessels (called arterioles),

Figure 32

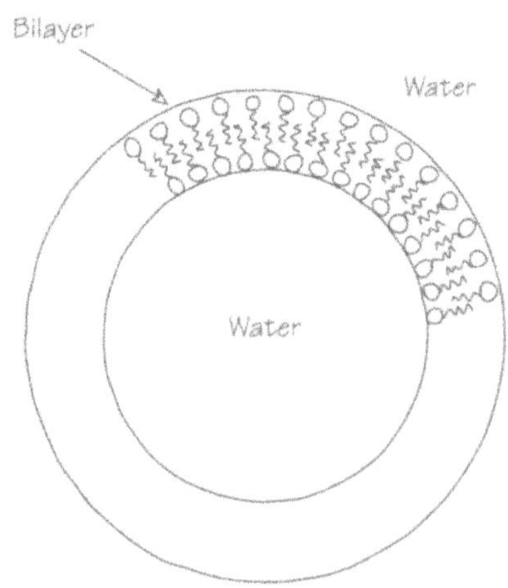

A bilayer folded back onto itself.

Figure 33

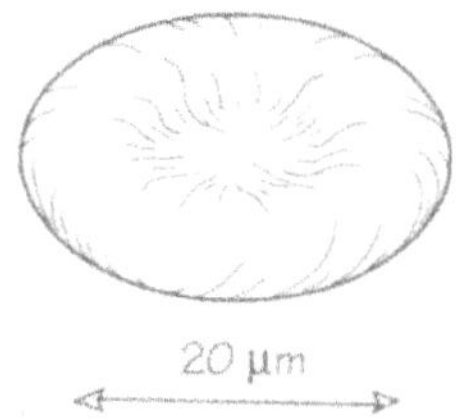

A red blood corpuscle is a bilayer.

often much narrower than itself. Every living cell is protected by a wall whose basic structure is that of a bilayer. Many high-school students know this, but very few people are aware of how the presence of a bilayer was first demonstrated. I learned this story quite recently during a seminar by Charles Tanford.

The idea is very similar to that which I described in connection with Benjamin Franklin. In 1925, two Dutch physicians, Gorter and Grindel, had the idea that the cellular wall, which is exposed to an aqueous medium both within and outside, must be made of bilayers. In order to prove this, they took a suspension of red cells and determined their total surface area by observations with an optical microscope. Next, with standard techniques, they extracted the lipids (the surfactants) present within the wall. By dividing the volume of the lipids by the surface area, they determined the thickness of the wall, which turned out to correspond to *two molecular lengths*. Here again, long before the advent of sophisticated tools (like the electron microscope), the Benjamin Franklin spirit scored another important victory.

Red blood cells have been studied for more than 200 years. At the beginning of the twentieth century, a curious property of theirs was noticed: they appear to twinkle like stars on a clear, dark night. Naturalists began to give free rein to their imagination, and some expressed the belief that this twinkling was some kind of manifestation of life. A somewhat obscure philosophy.

Our group at the IPC succeeded in demonstrating that it was, in fact, nothing of the sort. J.F. Lennon, a biologist, did some beautiful experiments on that scintillation process. He did not just observe the cell as a whole, but conducted what are called *correlation studies*. He observed the cell:

- at a particular point and at a given time, and
- at another point some time later.

He established how the scintillation varies according to the distance between the two points and the time interval between the observations. These kinds of studies are not easy from an optics point of view, but they provide a great deal of details concerning the behavior of the surface.

An excellent collaboration developed between Lennon and a theoretician from our team, F. Brochard. Ms. Brochard was able to show that the scintillation had nothing at all to do with a complicated message of life, but was merely evidence of thermal agitation.

In a liquid, molecules move about in all directions, quite fast. The higher the temperature, the faster their motion, and the more frequently they collide with one another. Molecules are too small to be observed directly in a microscope. It is, therefore, impossible to see directly either their motion or the thermal agitation which causes it. Placing a rigid particle a few microns in size (like our red blood cell) into the fluid does not "reveal" the thermal agitation either. Although the particle undergoes a large number of collisions with molecules every second, it does not move. For one thing, it is too heavy; for another, the innumerable collisions from all directions cancel one another's effects.

Our red blood corpuscle is too heavy to move as a whole under the influence of thermal agitation. Yet it makes an excellent indirect witness. The reason is that the bilayer which encloses it has a *very soft* surface. It takes very little energy to deform it, and molecular collisions have enough energy to do the job. It is these local and random deformations (Figure 34) which cause the twinkling: small sections of the surface act like so many independent oscillating mirrors: The scintillation is a manifestation of thermal agitation.

At about this time (1973–74), we sensed that these easily deformable surfaces, called random surfaces, constituted an interesting field of study.

Figure 34

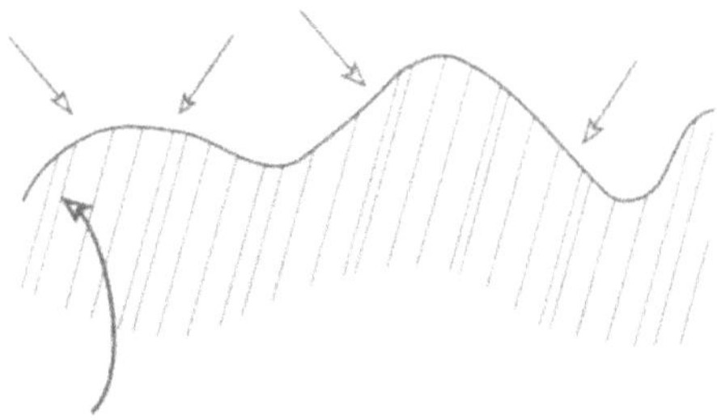

Random motion of the outer surface of a bilayer.

The field subsequently did develop considerably, and even had a few applications to the problem of contacts between living cells, contacts which play a considerable role in biology. W. Helfrich, the same individual who invented liquid crystal displays, was a notable contributor to this area as well.

The purely geometrical study of random surfaces has also been the object of much research. This science is beginning to bear fruits today in some emerging theories aimed at explaining the behavior of elementary particles—the ultimate constituents of matter. These can no longer be viewed as simple points endowed with usual properties, like mass and electrical charge. One is led to think of them as small rings (a little like smoke rings). In time, the ring evolves into a sort of a tube, which is a random surface. Here is a truly amazing concept, applicable on a scale of a few centimeters (for a smoke ring), as well as on a scale 10 billion times smaller (for contacts between living cells), and even another 10 billion times smaller still (for elementary particles). It is gratifying to realize that a rather mundane problem like the twinkling of red blood cells can be related to such a wide range of different phenomena. To appreciate how everything is connected in the realm of ideas is one of the great joys experienced by scientists.

CHAPTER 7

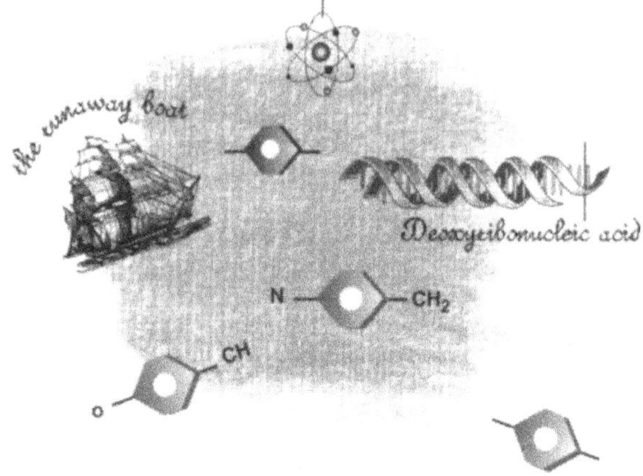

Bubbles and Foams

Soap bubbles are pretty, colored, mobile, and fragile. What a pleasure it is to show them to a child! During my lectures in high schools, I would often show slides of art reproductions, for example a Chardin[1] showing a child and a bubble, or a Japanese painting of the same period by an artist named Shibakoyan, in which an infant hangs on to his mother while trying to catch a bubble.

A soap bubble manifests every phase of life. It is born, it grows and develops, it ages, and finally . . . it disappears.

Every Color of the Rainbow

Bubbles are nothing more than a film of soapy water (Figure 35). Why do these films display colors? The answer is quite simple. Start by mak-

[1] Translator's note: Jean Baptiste Siméon Chardin, French painter (1699–1779).

Figure 35

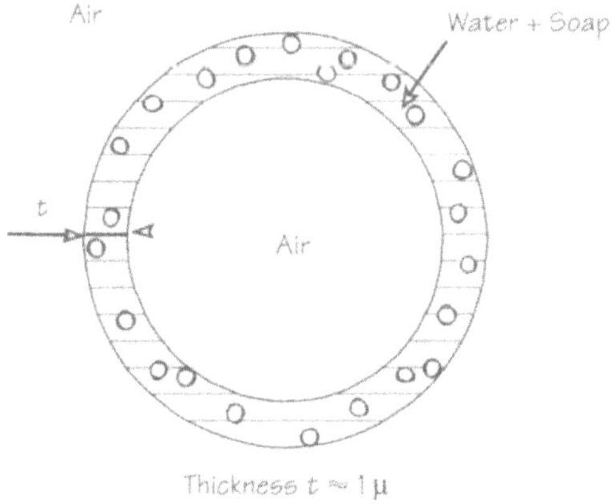

Structure of a soap bubble: Colored, fragile, and agitated.

ing a film yourself. You can use a wire egg whisk, or any loop with a handle, whose "scientific" form is a frame made of metal wire as shown in Figure 36. Immerse your frame in a solution of water and soap, and pull it out. The film will be relatively thick: a few microns, or a few millionths of a meter.

Figure 36

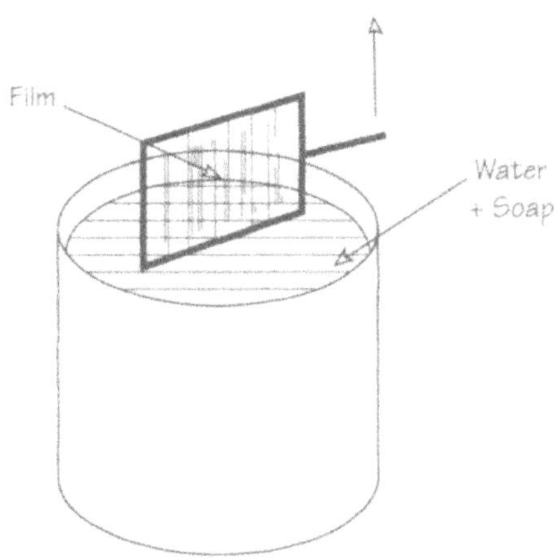

The time-honored art of making soap films.

Light waves reflected simultaneously by the front and back surfaces of the film interfere with one another, which means that they either add up or cancel each other depending on their wavelengths, relative to the thickness of the film. Because the different colors of the light illuminating the film correspond to different wavelengths, the colors are not all reflected in the same way. That is what gives the reflected light its color. For example, if white light illuminates a film of thickness such that the color yellow ($\lambda = 0.6$ micron) is canceled by destructive interference, only blue and red remain in the reflected light, and the film appears purple.

Newton was the first to understand these phenomena.

Water's Skin

Bubbles, soap films, that is, are colored. We have just seen why. The question now is: Why is it necessary to add soap to water in order to make bubbles?

The answer is that making a bubble amounts to forcing water to share a large surface area with air. But that is one thing water molecules hate to do. Each water molecule likes nothing more than to be surrounded by other water molecules. Those at the surface are cut off from half their neighbors. The energy required to create that surface is the same as that required to remove a layer (fictitious) of water molecules located above it (Figure 37). This energy, called the surface energy, turns out to be proportional to the surface area.

Figure 37

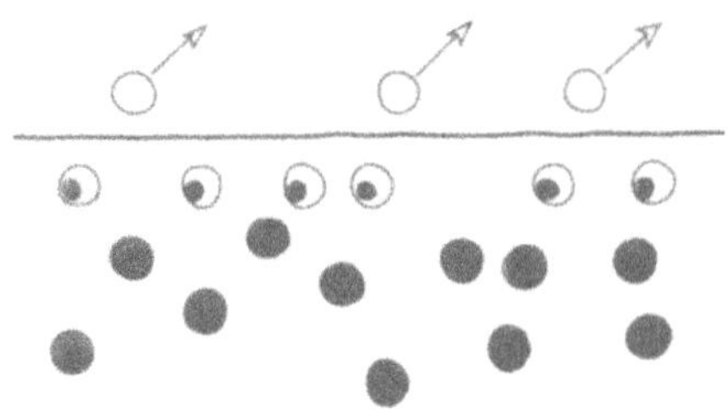

The "skin" of water.

Every physical system in nature seeks to minimize its energy. That is why the shaggy hairs of a dry paintbrush go limp and stick together after being dipped in water (Figure 38). As a true theoretician, I must warn you right away that a paintbrush with many hairs is too complicated an object. Let us, therefore, concentrate on a "theoretical" paintbrush with only two hairs. Once wetted, there is between the two hairs a film of water which wants to decrease its surface area. Hence, the water exerts on the hairs a force which tends to pull them toward each other. This force, equal to γl, is proportional to the wetted length l of each hair (this is just another way of saying that the force is proportional to the area). The proportionality coefficient γ is called the *surface tension.*

After Isaac Newton, who explained the origin of the color of bubbles, and Thomas Young, who understood the role of surface tension, I want to mention another great experimentalist in this area, J.A.F. Plateau. A particularly careful experimenter, Plateau one day made the following entry in his laboratory notebook: "From this day on forward, experiments recorded herein will no longer have been performed by myself, as I can no longer see well enough to conduct them." Though Plateau was going blind, he continued his work.

Returning to the topic at hand, we can say that a film always seeks to minimize its surface area. If one draws a film with a ring, one obtains a

Figure 38

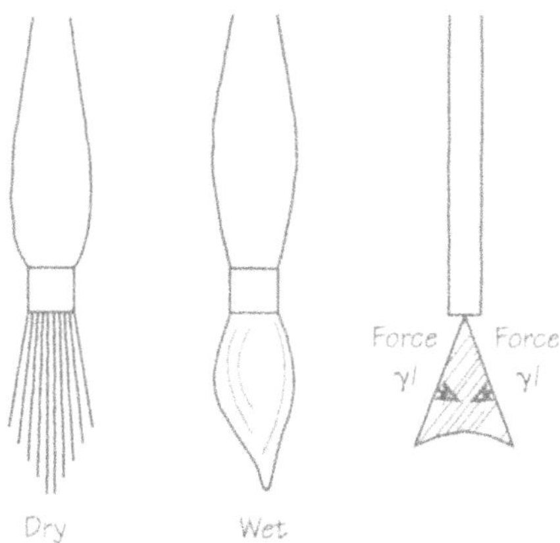

The wetted paintbrush.

Figure 39

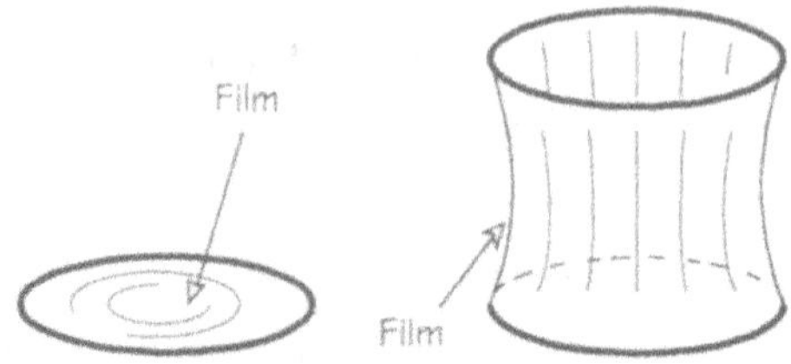

Minimal surface films.

flat disk. With two rings, the result is a surface which resembles the smokestacks of nuclear power plants (Figure 39). It was these types of films that Plateau worked with: what mathematicians call minimal surfaces.

SURFACTANTS DECREASE THE SURFACE ENERGY

Up to now, I have talked mainly of what might be called the skin of water, and of the energy associated with the area of the skin. But what is the role of soap in bubbles and in films?

Soap modifies surface energy. Just what is a soap molecule? It is one of those ambivalent molecules which we have already met in the previous chapter. A soap molecule has a hydrophilic polar head attached to a chain of 12 to 18 carbon atoms (with their peripheral hydrogen atoms). It is a molecule belonging to the family of surfactants (Figure 40).

Figure 40

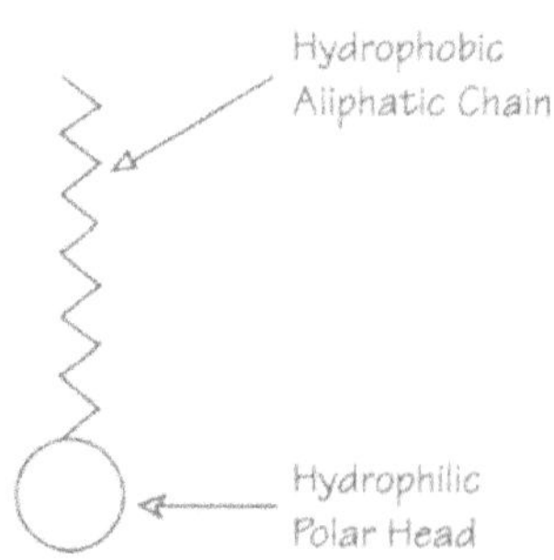

The "contradictory" parts of a molecule of surfactant.

Figure 41

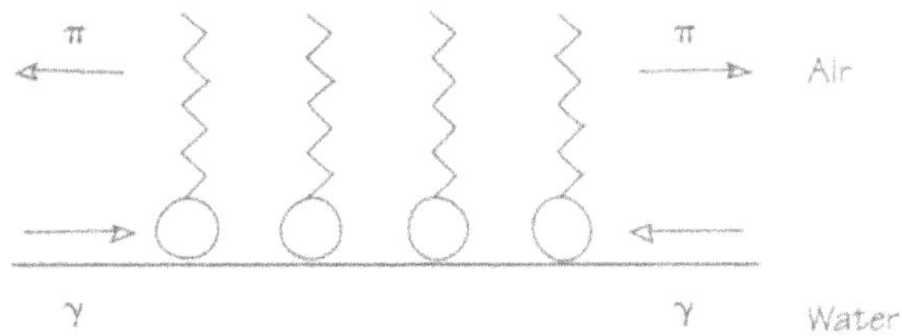

A monolayer of surfactant on the surface of water.

We have already seen that these molecules like to arrange themselves side by side. Their polar heads, which are hydrophilic, are immersed in water, and their hydrocarbon (or aliphatic) chains protrude outside the water, parallel to each other (Figure 41). An extraordinarily minute quantity of these molecules is enough to completely cover the surface of water with a single layer (a monolayer). About 20 milligrams is all it takes to cover 1 square meter of water. The molecules are "happiest" on the surface. In the language of physics, that means that the surface energy—the coefficient γ—is lowered (roughly by a factor of 10). In other words, the presence of a surfactant facilitates the formation of films with larger surface areas.

The Birth of a Film

The action of a surfactant is more subtle than the previous illustration suggests. Consider a film drawn with a frame from soapy water (Figure 42). On the basis of the color pattern, this film must be relatively thick at the bottom (some 10 microns), and thinner at the top (only a few microns). How does it hold together? The answer is not obvious. If the same experiment is attempted with pure water, the film will simply not hold, as we all know. Actually, the column of water has a serious problem: its own weight. It desperately wants to fall back down. How does it manage to resist this urge? The answer has to do with the monolayer of surfactant. The concentration of surfactants is dense near the foot of the column, and less dense toward the top. The surfactant molecules act somewhat like members of a crowd, pushing and elbowing each

Figure 42

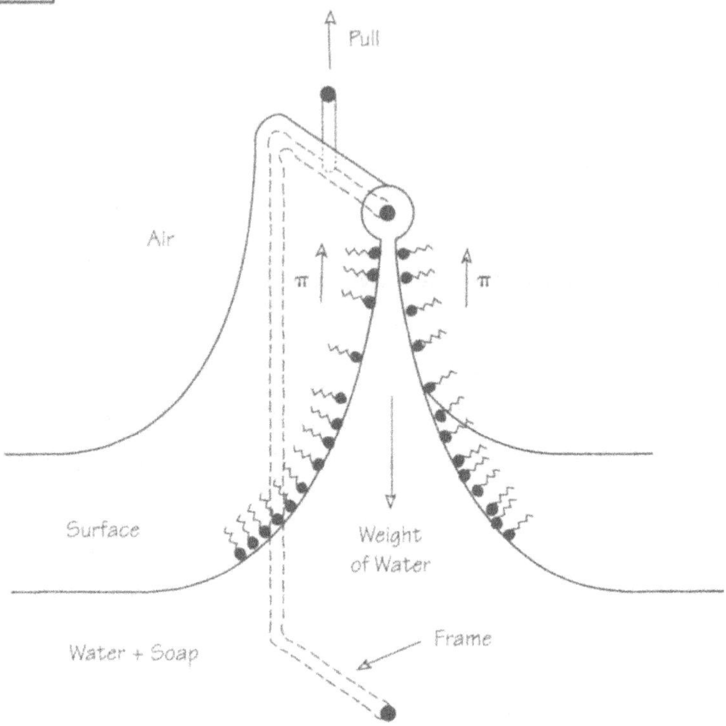

The physics of making a soap film.

other. They exert on the water a surface pressure π, studied in detail by I. Langmuir in the 1930s. This surface pressure is very strong at the bottom, less so at the top. Hence, the "bottom" effectively pushes the "top." We have a mechanism which opposes the weight of the column and ensures the integrity of the film.

It is interesting to explore the details of this mechanism and estimate the maximum height of a film that can be raised. Using the kinds of surfactant concentrations that can be introduced without problems in the solution, it is possible to extract a film a few meters high. That is the principle behind those circus tricks in which lovely ladies trap themselves inside large bubbles, roughly two meters in height.

A Self-Repairing Bandage

The surfactant has another intriguing and important property: it is "self-healing." J.W. Gibbs (1839–1903) was the first to observe this phenomenon, around 1890. Gibbs, one of the great theorists of the nineteenth century, founded the field of statistical physics. But he was also an active experimentalist, involved in the study of bubbles. He spent months staring at bubbles. Inspired by what he saw, he proceeded to develop what we call today the thermodynamics of interfaces. He managed to understand, among other things, an important property of films: self-healing.

If, for whatever reason, a hole forms in a film (Figure 43), the surfactant molecules, pushed by their neighbors, rush in to fill the gap. Imagine a crowd. During a ceremony, the police protect an empty space at the center of the square. At the end of the proceedings, the police withdraw and the crowd immediately invades that space.

This property, which imparts great stability to the film, is important in many other areas, for instance, in lubrication problems. In order to protect two metal surfaces in high-speed motion relative to each other, separated by a small gap, one traditionally uses fatty compounds. They too are surfactants, and so possess the property of *auto-cicatrization*, as self-healing is also called. The film they form experiences tremendous

Figure 43

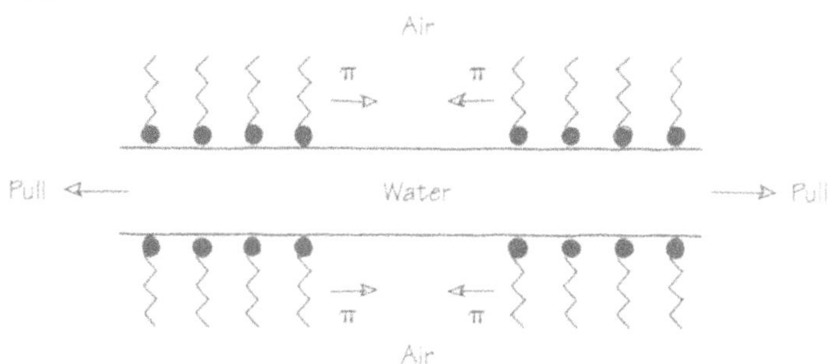

The self-healing characteristics of surfactant films.

stresses, as well as high temperatures. It is subjected to pressures of several hundreds of atmospheres. Under such harsh conditions, the film undergoes rapid chemical degradation, but fortunately this does not result in an all-out catastrophe, because fresh surfactant molecules spontaneously arrive at the proper spots to replace their degraded counterparts. This process is one of the marvels of lubrication.

For these times, marked by a concerted search for "intelligent" materials, we have here a beautiful example of a self-healing substance.

And So Goes Life

A bubble is born, grows, matures, ages.

The life cycle of films was studied by a great scientist named Karol Mysels, who in the 1950s wrote a remarkable book on the topic, illustrated by fabulous photographs. Karol, now retired in San Diego, California, continues to perform experiments of exquisite sensitivity. Before plunging into the life of films, I would like to take a moment to recount an anecdote taken from his laboratory notebook, the one in which Karol documented his observations. During one delicate experiment, he stopped and wrote down in his notebook: "At [such and such time], I interrupted my experiment. I observed a spurious signal. This signal did not come from my apparatus; it was completely external. I believe it was an earthquake." He had indeed detected an earthquake in San Francisco, at the other end of California. What a marvelous experimentalist!

Let us now return to our films, in their mature stage. In a mature film, drawn by a frame, *black zones* develop near the top and on the sides (Figure 44). In Newton's language, "black" is synonymous with "very thin": In these regions the thickness of the film is no longer sufficient for the reflections from the two surfaces to produce constructive interference, and the two reflected rays cancel each other exactly. Why do these very thin regions develop (Figure 45)?

Recall the van der Waals forces, which we have already encountered several times. They are attractive forces between molecules, effective over distances that are relatively long (about 10 nanometers) compared with the typical distance between two molecules (0.1 to 1 nanometer).

Figure 44

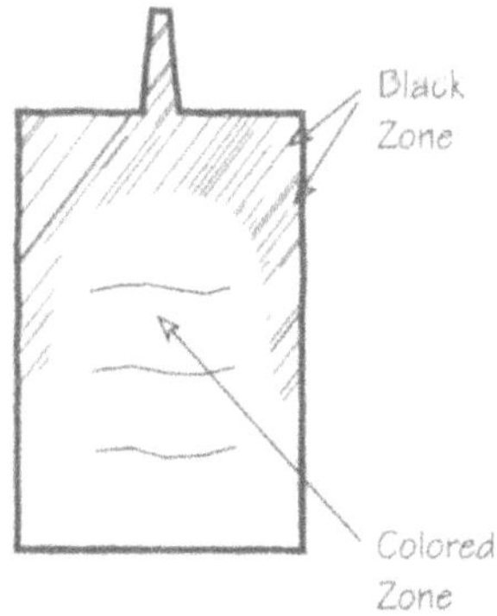

The different zones of a soap film.

Figure 45

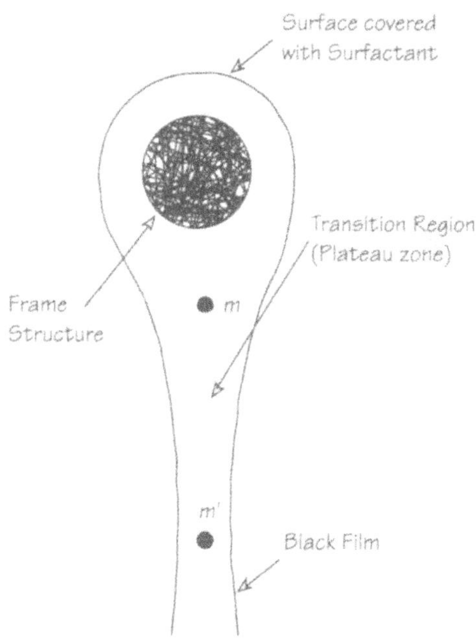

Structure of a soap film in the vicinity of the frame.

In the vicinity of the frame, the bulge of water is thicker than the film. A molecule, m, in the bulge is surrounded by neighbors, both near and far. It is happy. A molecule, m', in the film, by contrast, is still surrounded by near neighbors, but it has lost most of its far neighbors: it is not so happy. It rushes out to join the crowd in the more comfortable bulge, called Plateau's zone. In other words, the film empties itself into the thicker zone which is already well populated with molecules ("only the rich can get loans . . ."). The film gets thinner, and, in the process, turns black.

A Turbulent Bubble

I have mentioned an interesting property of bubbles: the extraordinary agitation of their surface.

Let us look at a colored film (Figure 46 and the cover of this book). Being colored, it must be thick, as Newton taught us. Occasionally, we observe a black (therefore very thin) blotch, M, which travels up toward the top of the frame. Why?

In a hot-air balloon, the air inside is lighter than cold air outside, and provides lift to the balloon. Here, we have a similar situation. The small black circle, M, is extremely thin. It contains very little water, and consequently is lighter than the thick and heavier colored regions which surround it. Like a hot-air balloon, it rises. The vertical portions of the frame in Figure 44 generate a black film and are the analog of warm zones: they produce a turbulent flow rising toward the upper part of the frame, much like warm air near a wall radiator.

It can be claimed (more pedantically, perhaps) that *turbulence* is easy to observe in a film. Here I cannot resist recalling another experiment in the Benjamin Franklin mold, one concerning the turbulence behind an object. If you look down the length of a bridge spanning a fairly fast river, you may observe behind each pier a peculiar and "turbulent" flow. This turbulence exists only if the river is deep enough to minimize the friction with the bottom. Soap films enhance turbulence because they are well away from solid boundaries. The bottom is not relevant, since the film covers only the surface of the water. Hence, any friction which

Figure 46

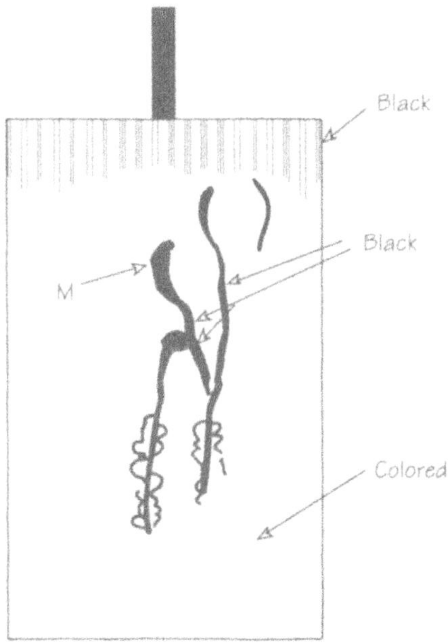

The dynamics of the structure of a soap film.

might dampen the turbulence is exceedingly weak, and the thin air above the film can do little to slow down the chaotic movements.

It is tempting to study these flows in the particular case of two dimensions (in a plane, or a film). This is precisely what was done by a great theoretician, E. Siggia, who used one of the first Cray supercomputers.[2] His calculations predicted an extraordinary turbulent regime. But a French mechanical engineer named Yves Couder played the role of a Benjamin Franklin in this matter. His experiment consisted in spreading a large film of soap (perhaps 1 meter long by 25 centimeters wide) on a horizontal frame and disturbing the film by drawing a rod through it. He moved the rod at a speed of a few centimeters per second (Figure 47). This experiment produces truly stunning phenomena.

[2] Cray is the trade name of the world's most powerful supercomputers, manufactured in the United States by Cray Research, Inc.

Figure 47

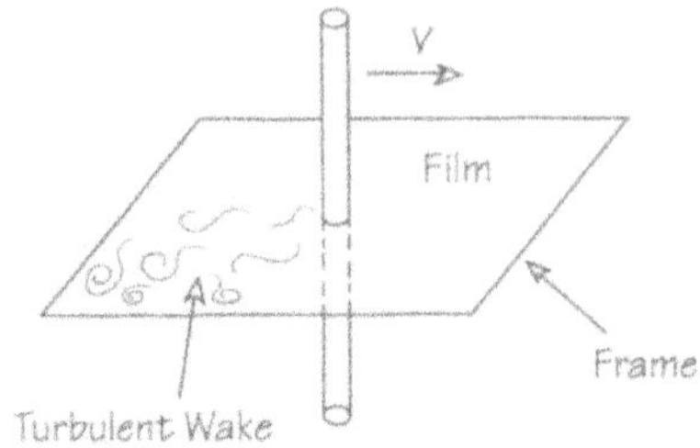

Planar turbulence created by an obstacle moved across a film.

When the rod is moved at a very low speed, the structure of the wake is simple and well known: behind the rod an eddy appears, then another. They form a fairly regular path, called a *von Karmann path*, like a row of trees along a road. But with higher speeds, the turbulence is stronger, and the path becomes chaotic. Figure 48, generously provided by Couder, shows the corresponding wake. The eddies react on each other. Occasionally they collide. Frequently, one observes pairs of eddies twirling in opposite directions. If nothing were to stop them, the pairs would go on indefinitely. But certain pairs collide, and the collisions can give rise to a host of chaotic events. The patterns observed are the same in style, detail, and correlation as those originally computed by Siggia, but the cost of Couder's equipment was less than $1000. (Needless to say, the cost of a Cray is on a different scale altogether.) Hydrodynamics does need numerical calculations. But it can also greatly benefit from the Franklin spirit displayed by Couder.

Newton and the Babylonian Sage

Who was the first to observe black films? For a long time, we all believed it was Newton. That documented account seemed ancient enough. As it turns out, though, the first reference to black films is found on a Babylonian tablet nearly four thousand years old. Moreover, this tablet is not unique. There exists a whole collection of similar tablets at the Collège de France (Figure 49).

Figure 48

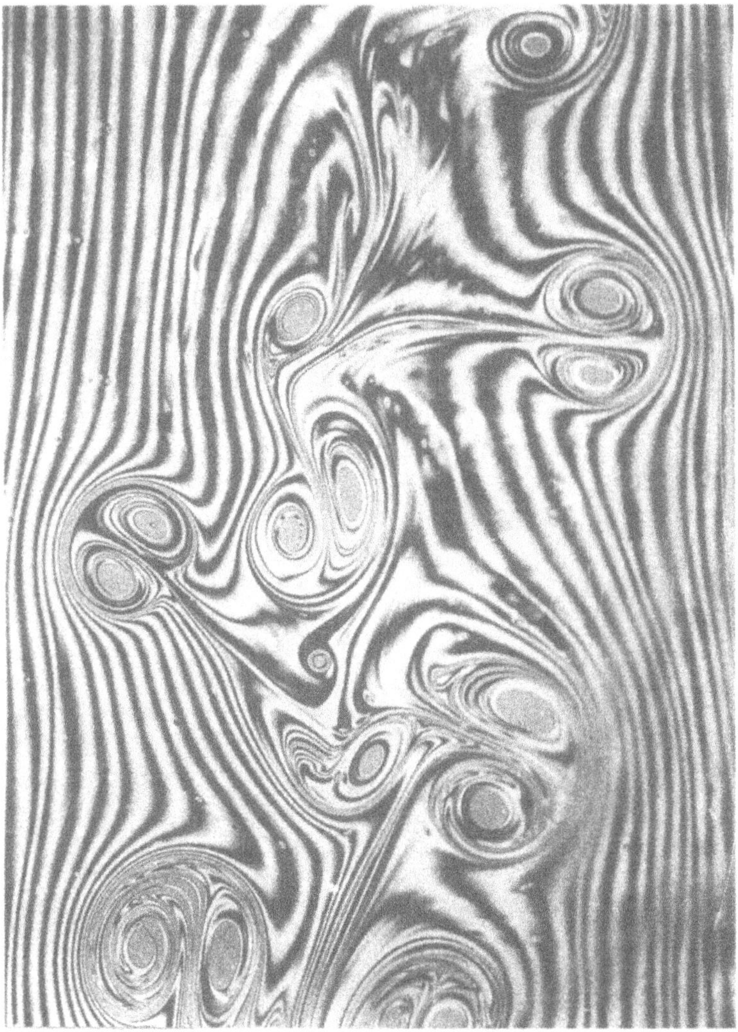

Turbulent pattern observed by Y. Couder. (Courtesy Y. Couder.)

An important endeavor for the Babylonians (as it is for us) was to try to divine the future. They explored every method, including the most elaborate ones, to predict future events. In modern times, it could be claimed that supercomputers are one means of guessing the future—in meteorology, for example. Tea leaves and coffee grounds have also

Figure 49

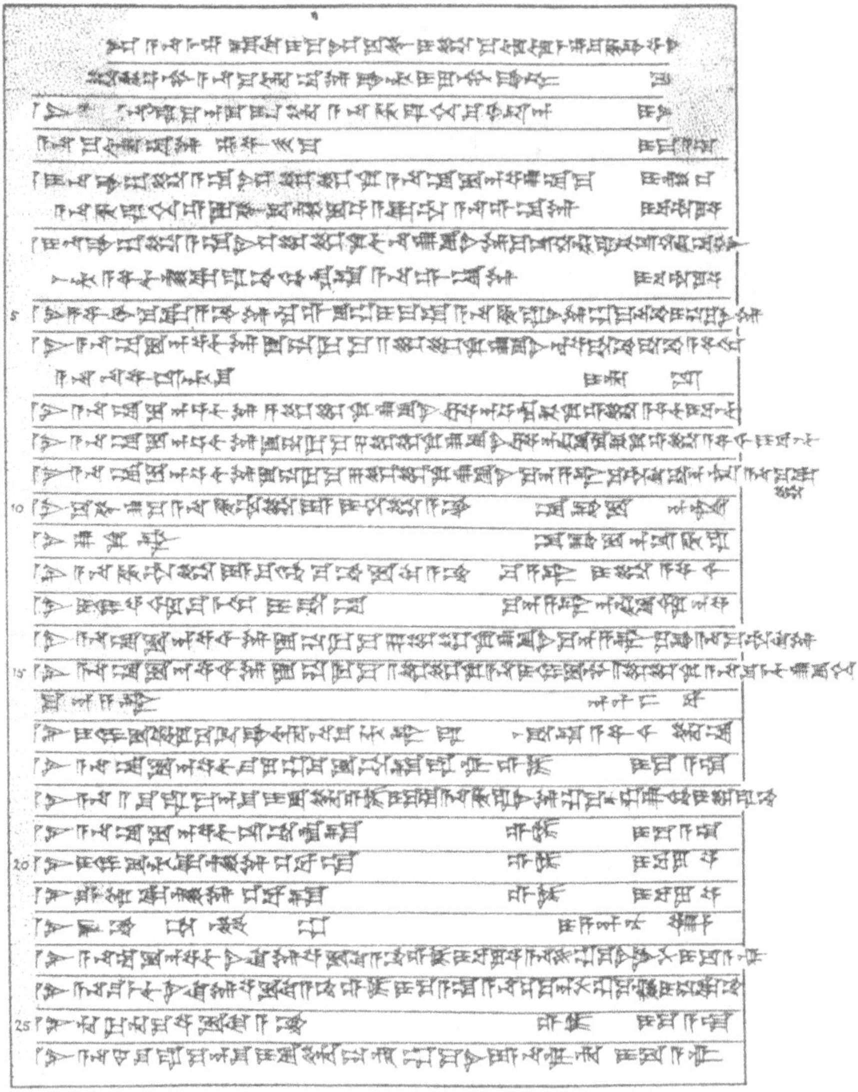

The reference to black film made by a Babylonian sage—a perfectly lucid description . . . (Reproduced from *Cuneiform Text from Babylonian Tablets in the British Museum,* vol. 5, London, 1898. Communicated by the Department of Assyriology, Collège de France, Paris.)

gained a certain reputation as having fortune telling powers. They have the great advantage of being a lot cheaper.

Most divining methods utilize highly unstable systems, giving rise to naturally chaotic situations. The Assyrians settled on a bowl containing water and oil. By shaking it, they created films. They often witnessed the appearance of black regions, whose behavior they would try to interpret. Imagine the responsibility of a high priest granting a couple an audience. Should they marry or not? The sage looks for two black spots and observes whether they unite, separate—or worse, get devoured by a larger black spot. This divining technique is named *lecanomancy* (the prefix *lecano*-means "bowl"). The texts that describe it are very moving: they give us the very first recorded description of an object as fragile and as remarkable as a black film.

Rose, She Lived the Way Roses Do

We all know that a bubble bursts suddenly. On the scale of our visual perception (1/30th of a second), we can see nothing at all. It is sudden death.

Karol Mysels did some beautiful experiments on the destruction of bubbles, which he triggered by causing a spark to jump between two electrodes placed on either side of the film (Figure 50). The spark initiates a tiny hole which grows rapidly through the action of the surface tension γ. A bulge forms at the periphery of the hole, collecting water

Figure 50

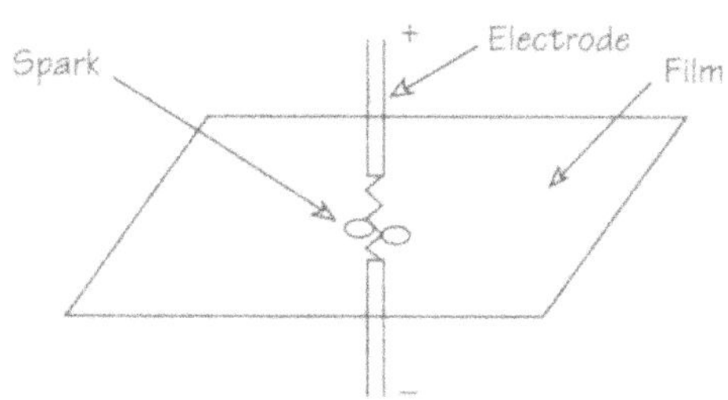

Triggering the death of a soap film with an electric spark.

Figure 51

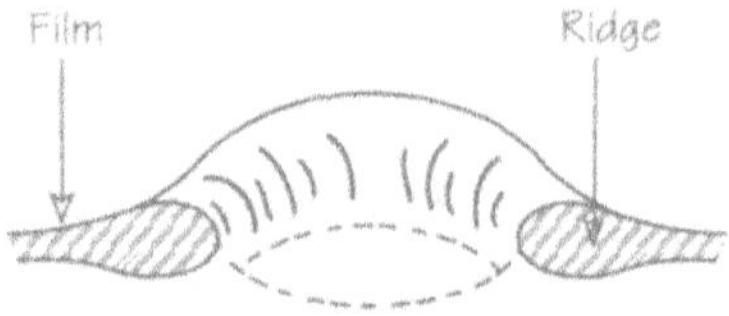

Structure of the expanding hole preceding the death of a bubble.

as the hole expands (Figure 51). The speed of growth of the radius R of the hole turns out to be a constant, on the order of 10 meters per second. It is easy to photograph such phenomena with a flash lasting only 1 microsecond (a bubble 10 centimeters in diameter bursts in $\frac{1}{100}$th of a second).

Drowning by Numbers—or One Bubble, Two Bubbles . . . Foam

The study of bubbles is obviously directly related to that of foams. Indeed, foam can be viewed as a collection of a large number of bubbles. Foams play a considerable role in our practical lives. It is important to know how to create them, as well as how to eliminate them.

While we can all appreciate the benefits of foam in a shampoo, the same kind of foam should be carefully avoided in washing machines or dishwashers. Detergent manufacturers are ever alert to this risk and take precautions to spare us the misadventure of the sorcerer's apprentice: a wave of foam inexorably invading first our machine, then our kitchen. The chemical industry sometimes runs into similar problems: huge reactors, columns 20 or 30 meters high, overrun by a pesky foam—a veritable catastrophe.

A few years ago, wastewater was routinely released back into rivers. It was not uncommon to see huge mountains of foam at the point of discharge. Today, we have access to antifoam compounds that can break these films. These compounds are seemingly magical products: one drop in a large container of foam, and the bubbles disappear before your eyes.

In some cases, foams do serve a useful purpose. For instance, they can create a temporary barrier between a fire and an oil field. Many fire extinguishers contain a foam.

It must be emphasized that consumer products, including myriad soaps, detergents, and lathers, do not, by a long shot, represent the most important market for the surfactant industry, which manufactures about 10 kilograms (20 pounds) of surfactant for every man, woman, and child on this planet, every year. Most of this production is used in the extraction of metals from low-yield ores. We are increasingly being forced to exploit ores of low content (sometimes around 1 percent), which would have been snubbed not long ago. In finely ground ore, the metal particles (often in the form of oxides) are swept up by the bubbles of a foam, making them easy to separate from the tailings. This process is known as flotation.

Questions and Answers

Foam and the Edges of the Universe

Of all the questions asked after this lecture, I most remember the following one, far removed from our mundane world.

Q.: You have studied matter on an infinitely small scale. It would seem that this would force you to consider the infinitely large as well. Would you care to share with us your concept of the universe?

A.: No, because I lack a solid formation as an astrophysicist. I simply cannot answer this question credibly. But I can provide an answer on a smaller point which, to some extent, is related to the sciences that I have just described to you.

Based on a number of observations, many people believe that, at time zero, the universe was infinitely dense. This system expanded and evolved toward lower densities and lower temperatures. We finally have today a relatively frozen universe which offers some clues as to how things were, perhaps a few short milliseconds after the Big Bang. There are two aspects to this situation which are decidedly contradictory. The thermal infrared radiation left over from that time is perfectly isotropic; it is the same in all directions of the sky. On the other hand, the distribution of faraway galaxies is not at all uniform in the sky. It is, actually, somewhat analogous to a foam; it has what is called a fractal structure.

As it happens, these fractal structures are typical of cooperative phenomena that occur in phase transitions. In the transition between the vapor phase and the liquid phase of water, there exists a particularly interesting regime when the vapor has roughly the same density as the liquid, which occurs at the so-called critical point. If, near this point, one examines things on a very small scale, one observes fractal structures in the arrangement of the microdroplets of vapor and of liquid.

The fact that somewhat similar structures show up on very large scales suggests that the universe, perhaps at a very early age in its life, underwent a process analogous to a phase transition. It is the remnants of this transition that we may be looking at today in the distribution of galaxies.

CHAPTER 8

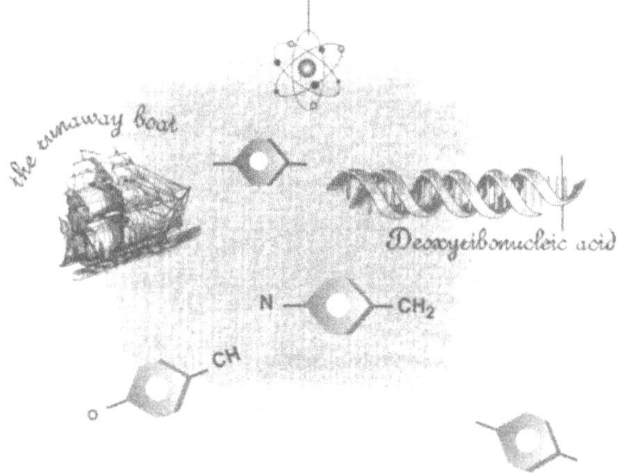

Fragile Objects

A little Arabic gum stabilizes Chinese ink for a full year. A few drops of shampoo generate a cloud of lather. Traces of oxygen in the air transform hevea sap into rubber. A tiny battery is enough to update the display of a watch. Every one of these cases epitomizes the essence of soft matter. Soft matter is exactly like the clay of the sculptor, which, through an extremely weak action, can be changed, shaped, and organized.

And, also like clay, the result itself is often an object that is fragile and deformable: for instance, rubber, half-way between a solid and a liquid; foam; the red blood cell snaking its way through our smallest capillaries; the textile fiber extracted through a nozzle at 6000 meters per minute.

Our species initially learned to work hard objects, like flint, bronze, stone, brick, or even wood. But soon, it found itself in need of more nuances, of more pliant materials: leathers, natural fibers, waxes, starches—the science of caulking may well have initiated Phoenician navigation. Likewise, twentieth-century physics first devoted itself to hard materials, such as metals, semiconductors (which opened the way to our modern forms of communication), and, later, ceramics. But the recent trend has been in the direction of soft materials, of which polymers, detergents, and liquid crystals are the most common forms around us.

There can be no life without soft matter. Every biological structure (the molecules containing the genetic code, proteins, membranes) is founded on that very concept. But it is imperative to soften the almost imperialistic undertone of such a sweeping statement. Never fall in the trap of believing, as did Auguste Comte, that physics is destined to guide biology. Physics can only provide a general framework. Biology has its own methods of observation and discovery. Take, for instance, the example of adhesive molecules, responsible for holding neighboring cells together by establishing selective links between them. These are rather complicated proteins, attached to membranes. A physicist would be likely to maintain that, in order be understood, such molecules must be isolated, crystallized, and their structure studied with X-rays. But programs to accomplish this are painfully slow, almost clumsy. Biologists, on the other hand, through a clever use of molecular genetics, quickly determined the chemical sequence of each chain, and, from there, were able to group molecules into families. They established that in each family there is a "tail" buried inside the cell, a "waist" embedded in the membrane, and a complex "head" able to recognize certain external partners. Through this biological approach, the shape of the head and what it does are beginning to be understood long before the first salvo from the physicist's heavy artillery.

Living matter is based on the principles of soft matter, but with a subtlety which is often beyond the realm of physicists. It is, in fact, the reverse link which becomes intriguing: biological systems, once elucidated, give us fresh ideas to invent new objects. Consider, for instance, the bones of vertebrates: they are marvelous structures which have the ability to repair themselves. Is it possible to conceive of a synthetic material with the same property? Progress is actually being made on that front: it is already possible to fabricate structures capable of experiencing something analogous to pain when they are "mistreated." It is done with a fiber-optic network, which detects the presence of a deformed region, measures the magnitude of the damage, and pinpoints the location of the affected area. The second phase is the repair process. Here, we are not much beyond the babbling stage. Nevertheless, one can conceptualize a network of capillaries containing suitable reagents: should the capillaries break in an accident, the reagents would be liberated and could start synthesizing new polymers on the spot. The technology of these hypothetical structures mimicking living matter will benefit immensely from progress in the field of soft matter.

In summary, fragile objects are an important element of man's future technologies, as well as an essential foundation of life itself. But what we hope most to have conveyed in these pages is their contribution on a cultural level. Their science is one of practice, of finesse, of the manipulation of opposites. Recall, once again, the case of liquid crystals, where the challenge is to meld into a single molecule parts that are hard and parts that are soft. It is when dealing with soft matter that the Benjamin Franklin spirit, so often invoked here, manifests itself the most vividly—in the mind set that grasps the hidden simplicity of things, in the spirit of show and tell, in the tradition of the schoolteachers of the time of Jules Ferry, in the ability to marvel at a lowly drop of water.

PART II

Research

CHAPTER 1

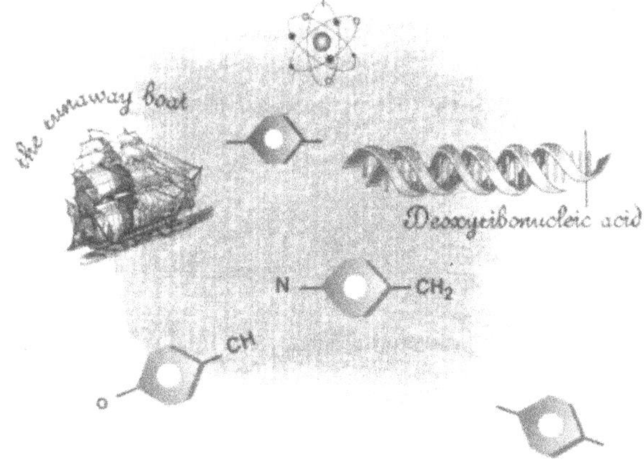

Profession: Research Scientist

Traditional wisdom would have us believe that scientific research is the purview of a small, mathematically gifted elite with a unique psychological profile. Nothing is further from the truth. A researcher cannot be defined in terms of a particular cultural mold. In particular, mathematics is not a universal prerequisite for scientific genius, contrary to the naive cliché of a Mr. Einstein seated in front of a blank piece of paper, suddenly scribbling $E = mc^2$, or of a Mr. Newton under his apple tree. Modern science does not at all conform to these popular myths. It is not the exclusive province of child prodigies. In physics, a very young researcher is rarely productive. When I first started working in laboratories, I produced nothing original or personal. It was only relatively late in my career, when I was working as an engineer with the Atomic Energy Center at Saclay, that I began to be consulted about strange effects in isotope separation, and that I felt bold enough to venture proposing some answers. By then, I was already 24 years old.

Does research imply a calling? It is difficult to tell. In my case, at age 16, I was just as much attracted by literary and artistic matters as I was by the sciences. Why did I eventually gamble in favor of the so-called exact sciences? Perhaps I was aware that in the sciences fast talkers and con artists tend to be eliminated a little more quickly. Perhaps I was guided by a love for a craft in the sense of the word meant by Claude Lévi-Strauss when he speaks of the "lost craft" in the art of painting. Modern painting is no longer a craft in the strictest sense, as it used to be in the days of the great masters of Italian Renaissance or of the Flemish school. Everything has become fair game. This is not so in the sciences and more generally in technology. When a part is being crafted in the machine shop, it must be "within specs," which means that it must conform to a high level of precision. The same type of philosophy applies to theoretical physics as well. That is not to say that the part produced is absolutely perfect, merely that it is the best that can be achieved given the present state of the art; it is impossible to cheat or equivocate.

There is another important difference: while literary and art criticism constitute legitimate professions, there is, fortunately, no equivalent in the sciences. This is significant. An artist often spends his entire life in a state of complete doubt as to the value of his work. In this respect, the sciences today are more relaxed: One can gauge more serenely the utility of one's work. This relative serenity is largely due to day-to-day activities spent interacting with a *group*, rather than as a lone contributor. Scientists live in a world ruled by collaboration. Teams are made up of individuals with diverse temperaments, scientific backgrounds, and aptitudes. They thrive on this very diversity and complementarity. Teamwork is one of the great advances of our time. Even in the nineteenth century, scientists like Young, Faraday, Rayleigh, or Ampère went through their entire career as solitary workers.

Teamwork

In a science as much in its infancy as soft matter, our current field of research, groups are typically quite small, composed of no more than four or five individuals—for example a chemist, an optical physicist, perhaps a condensed matter physicist, an electronics engineer, and, in a supporting role, a would-be theoretician. The more contrasts in the group, the better it performs. When a problem crops up that seems to stump everyone, one member is always ready to stimulate the others and to spark a discussion.

Small groups promote a special mood, not unlike that around a table at a tavern. Without the feedback and support of the team members, researchers would likely be driven to neurosis by the fierce competition both at the fundamental and applied levels. When a researcher is on a hot trail, he knows that several teams throughout the world are racing toward the same goal. Will he be able to come up in time with a new idea, a new instrument, a new observation, a new process? Or will he be beaten to the wire? Should he find himself overtaken two or three times by more powerful, better equipped labs, our researcher is at risk of throwing in the towel. "I will forever be beaten, I give up." One of the most effective antidotes is teamwork. Individuals benefit from the support of their co-workers.

At age 28, although a theoretician, I decided with remarkable innocence and naiveté to direct a group of experimentalists in superconductivity. It was a bit crazy. I knew very little about experimentation. Besides, this was a time when superconductivity was a hot topic in research. There was a great rush into the field, including the heavyweights, large companies like Bell Labs, RCA, Philips, and General Electric. And there we were, novices in the field, trying to contribute something new. How did we manage to survive? We did it by closing ranks and helping each other. We put all our eggs in *one* basket, one subject, instead of dividing our resources into four, five, or six theses, as was customary. Six individuals working on a single topic. After a year, when that topic had sufficiently progressed, we moved to another, and so on. This strategy kept us on a competitive level against the powerhouses.

Competition varies from one discipline to another. Consider the case of particle physics, which requires large accelerators. This is a discipline marked by exceptionally fierce rivalry. There are only three or four facilities in the world engaged in this type of research. When you run an experiment on one of these large machines, you are likely to be buzzed by a crowd. While you are struggling with your experiment, people mill about, hinting that you are wasting valuable machine time with no useful results, and that others could put it to better use to advance the cause of science. Working under such conditions is a painful ordeal. In our smaller sciences, the pressure on the playing field is not as intense.

Publish or Perish

The difficulties dogging the life of a researcher change from country to country. At the moment, France seems to me to be relatively shielded (at least as long as the government continues to maintain the course followed since the end of World War II). In the United States, on the other hand, the physical sciences are in the midst of a rather tense, if not critical, situation. Competition there is particularly harsh. To achieve employment security in the form of a tenured university post is an arduous trial. Researchers devote a large portion of their time building up material to gain promotions and to obtain funding. A perverse effect of this frantic competition is that young principal investigators tend to recruit an excessive number of doctoral candidates in order to turn out more publications and to secure more recognition. Some take on as many as 10 students—too large a group, honestly, to train properly. Programs are conducted too hurriedly and are vulnerable to shifts in fashionable topics.

The United States would be well advised to take a look at our National Center for Scientific Research (French acronym CNRS), where scientists enjoy permanent positions. This security allows them to conduct a more balanced, but also bolder, type of research.

The Two Halves of the Sky

As difficult to stereotype as they may be, do researchers still constitute a separate breed? Quite apart from specialization and the division of tasks, the fact that this community is so diverse does not help in trying to define it on a global level. Progress in science depends on both the fundamental and the applied sides of research, which are two rather different worlds, as we saw in Chapter 2 of Part 1. While science of course encompasses the entire spectrum from the most purely fundamental to the most extremely applied, in practice a certain chasm exists between academic or university scientists and industrial researchers. Although the situation has much improved of late, it may be useful to review how this state of affairs came about.

University researchers have traditionally had a rather negative view of their counterparts in industry. In my younger years, engineers were perceived as being rich in equipment and a little short on intellect. It was an unfortunate flaw in the French mentality to view fundamental research as more noble and more intelligent. The truth is that many products emanating from industrial research attest to an inventive talent at least as great as that of fundamental groups.

I will highlight one example among hundreds of the imagination of the industrial side of the house. To modify an interface between water and oil, the conventional approach is to resort to detergents, which, as we have learned, are small molecules whose head has an affinity toward water and whose tail has an affinity toward oil. At the Collège de France, we revisited this issue and tried to make not molecules but grains, solid particles, which we named *Janus grains* after a Roman deity often depicted with two faces in opposite directions. These grains have one side which likes water and the other which abhors it. C. Casagrande, the technological good fairy in our group, succeeded in obtaining grains with these contradictory properties, but with a technique that was rather primitive. The recipe consisted in taking glass marbles, depositing them on the surface of a liquid varnish, and making sure that they were only half immersed. Next, the exposed half was treated chemically to make it hydrophobic and the varnish was subsequently dissolved away. The process produced marbles which had peculiar properties at interfaces.

But we could not produce large quantities of those, just a few milligrams at a time. A German firm, Goldschmidt, devised a much smarter manufacturing technique. They started with an *aerogel*, a common industrial product which consists of hollow glass particles. The trick was to treat the external surface of the particles to render them hydrophobic (without affecting the internal surface), and then to break them in half to obtain small marbles, one side of which was hydrophilic and the other hydrophobic. With this process, it became possible to produce literally tons of Janus grains at a very low cost. I was quite impressed by the inventiveness of the Goldschmidt researchers.

In France, heavy industry has important research laboratories. But too often our corporations do not have the resources to afford research on long-term applications. The scientists are frequently overwhelmed by short-term problems. Their work consists primarily in supporting existing products and in improving them in the face of competition. They have to respond to the boss who announces: "The Japanese have just come out with a capacitor which is thinner than ours (or which costs a penny less apiece); if we fail to improve our process within the next six months, we will lose market share and be forced out of business." In this respect, the life of industrial researchers is undoubtedly more stressful than that of their cousins in the fundamental arena.

A Promising Formula: Mixed Laboratories

When a student asks himself if he should join a fundamental research lab rather than an industrial organization, my first reaction is to dedramatize the dilemma. Many decisions are not as momentous as they appear. What matters in life is less the contract itself than the spirit in which one decides to honor it—very much as it is in a marriage.

I know some researchers who were trained in a fundamental environment before gravitating toward industry. Some of my former students have become industrial leaders. Conversely, the greatest polymer theorist, Paul Flory, spent the first part of his career in industry. Nothing is irreversible. Besides, contacts between the two sides have proliferated in recent times. Mixed laboratories illustrate the point vividly, even if they involve only large industry. There is an excellent collaboration between Saint-Gobain and the National Center for Scientific Research in the area of glass surfaces, another between Rhône-Poulenc and the Atomic Energy Commission dealing with ultra finely divided matter.[1] These small groups typically involve five university researchers, five industrial engineers, five doctoral students, and a few technicians, and their work is beneficial to both camps. The mixed laboratory approach brings to the industrial researchers a lot of information that their academic co-workers are better able to collect, the former being held to more discretion because of trade secrets. The collaboration can open up some potentially fruitful avenues. Conversely, it benefits the fundamental side by virtue of the questions brought by industry, questions which are often profound and insightful for a fundamentalist. For a student sitting on the fence, the mixed laboratory can offer the ideal compromise. But, once again, the terms of the choice must be viewed in their proper perspective. Applied research is neither air-tight nor monolithic.

The High-Wire Dancer

In my small institute of higher education, the Institute of Physics and Chemistry of the City of Paris (affectionately known as PC, for short), roughly eighteen laboratories are engaged in research, which is often applied and is largely supported by industry. But we are not talking about just any kind of applied research. Those groups that get involved in experiments that are too commonplace, too repetitive, on systems made and manufactured by industry, are reined in. Conversely, those that stray in too fundamental a direction are gently but firmly pushed out of their ivory tower. I do not view it as healthy to do art for the sake of art when the research is subsidized by the taxpayer. At the same time, I recognize the risks of the researcher obsessed by "short-term" applications. To work too closely to technology can be stifling. And so, as director of the Institute, I spend a great deal of time lecturing my staff and making sure that nobody confuses the mission of a researcher with that of a development engineer—equally honorable missions, but not interchangeable.

---

[1] Translator's note: The Atomic Energy Commission is a government entity, known by its French acronym CEA, that is similar to the U.S. Department of Energy.

The origin of the Institute of Physics and Chemistry of Paris goes back to the war of 1870. The war was lost and, as a result, so were Alsace and Lorraine, including the city of Mulhouse. In Mulhouse had been the only French school of chemistry in which students were exposed to practical laboratory sessions that were directly linked to the theory taught in class. To remedy that loss, the Paris Municipal Council decided in 1880 to create a city-chartered school, with two primary mandates:

- Both physics and chemistry were to be taught on an equal footing.
- The students were to perform laboratory experiments based on the material presented in formal lectures.

These two guiding principles were settled upon during the remarkable deliberations of the City Council. The project got off to a quick start thanks to the availability of Alsatian refugees with a practical bent owing much to the German style of teaching. They had the wisdom to recruit Pierre Curie, then 25 years old. Very quickly, the Institute began to attract a number of enterprising students: among them were Pierre Langevin, theoretician in statistical physics, as well as the inventor of sonar; and Georges Claude, founder of the manufacturing company L'Air Liquide; and others.[2] And now it is our turn, with our modest means, to try to keep the tradition alive.

Currently, our Institute plays a leading role in widely different fields: the metallurgy of polymers, acoustics, electronics, and even the theory of soft matter. More traditional disciplines, like chemical synthesis, solid state physics, optics, have also managed to find new directions to explore.

We have in our Institute true inventors, like Jacques Lewiner in electronics, George Charpak in bioengineering, and many others.[3] One of

[2] Translator's note: L'Air Liquide is a large manufacturer of liquified air and other cryogenics.

[3] Translator's note: Charpak received the 1992 Nobel Prize in Physics, for his work in the development of the cloud chamber used to detect elementary particles.

my great satisfactions following the Nobel Prize was the opportunity to address Parisians and to convey to them the sense that the money they invest in this school is well spent.

My function is somewhat similar to that of Mr. Loyal, the circus character in Molière's Tartuffe, who follows the steps of a high-wire dancer as she performs her routine. His purpose is to prevent a fall on either side of the wire. Avoid straying toward either the extremely applied or the excessively fundamental. I must encourage and guide the dancer—a little to the left, a little to the right—without ever startling her.

Without a doubt, the desire to pursue a fundamental problem, without any concern for possible applications, is sometimes perfectly justified. Such is the case with some astrophysicists or some high-energy physicists. Scientists who slow down atoms with optical beams and cool them to temperatures as low as ten millionths of an absolute degree seem to be engaged in pure scientific play. Nevertheless, I believe that this kind of research deserves to be supported. It would be absurd to place scientific culture at the exclusive service of immediate applications, at least in Western societies.

In our own disciplines, the dilemma is ever present. One can tackle a beautiful problem without any clue as to where it will lead. Often nature is kind enough to come to our rescue by throwing bridges between the fundamental and the applied. A case in point was our work in the field of dewetting, which led to unexpected payoffs, for example, in the technology of high-speed magazine printing.

I am always much relieved when I can steer a young researcher toward a topic that has an impact on the real world, thereby preserving for him a greater range of future career options. Not very many scientists, nowadays, can be at once theorist and experimenter. A few do manage the challenge quite well, but they remain very much in the minority. More common, it seems to me, is the scientist capable of both devising an experiment and grasping the broad outlines of its interpretation.

This brings us back to the need for balance. Nothing, it seems to me, is more beneficial in life than the ability to correct one's weak points. The theorist who likes to ponder endlessly in his corner should be invited once in a while to come out and work with his hands. Likewise, the tinkerer should be persuaded to occasionally take the time to read works of general culture.

One Length Ahead

I must emphasize once again that research is not clearly segregated between fundamental and applied. To blend the two mind-sets is not only desirable, it is essential to maintaining our economic and industrial competitiveness. Some in the academic world are concerned when they see researchers stray toward the applied—presumably sacrificing their values—and sell themselves to industry. They express a point of view that gained popularity in France after the turmoil of 1968. It became the ultimate aspiration then to get involved in totally useless endeavors, without any practical consideration whatsoever. This kind of scientific dandyism has since somewhat regressed thanks to a rapprochement between the academic and industrial communities, which has fostered a better mutual understanding. But be ever mindful of the high-wire dancer! These contacts are quite necessary, but not to the point that research as a whole should become a satellite of industry. While we do train at the Institute of Physics and Chemistry students oriented toward industry, it is not to blindly follow established practices, but rather to contribute novel possibilities.

Research engineers who use the most complex and advanced tools and apply them to the real world find themselves naturally at the very nexus of the system. I myself have been employed by General Electric, Philips, and Exxon, and I currently work at Rhône-Poulenc—all without the formal title of engineer.

Before addressing an audience of university graduates, I once looked up the origin of the word *engineer* and its changing definition. In the twelfth century the term was used to designate an "ingenious" (the word has the same root) individual who fabricated machines, particularly of the military type, and this meaning was to survive for a long time. But if one follows the evolution of the craft, one finds that it matches closely that of our civilization. It ushers in and accompanies the march of modern times. If Western nations still enjoy any relative supremacy at all, it is precisely because, from the Renaissance on, the engineer began to engage in increasingly broader activities, first in Italy, and then in the Western world at large. It is important to remind the younger generations of this fact: they must be aware that the advantage of the West, which they take for granted, will not survive long if we fail to maintain the spirit of research and innovation that has permitted us to pull ahead by a length.

In the original definitions of the trade of engineer, the word *inventor* is included right alongside *builder.* Yet this inventiveness is all too often neglected in the training of engineers, who are frequently prone to drifting toward administrative and commercial management, or toward an essentially passive function of production. Without the influx of Chinese and other immigrants, American scientific research would probably be in danger. Should our own system in France evolve in the same way, we would surely be heading in the direction of some third world countries, which, not motivated to invent new processes, simply buy them, exploit them under license, and manage factories.

The dark period which the Swiss watch industry has just gone through demonstrates the vulnerability of any industrial sector that ignores long-term research. That industry believed that it was perfectly safe from competition. So sure was it of its tradition of excellence that it failed to foresee the possibilities of liquid crystal displays (LCDs). The result of this blind confidence was that traditional watch manufacturers took a beating on a massive scale. Not until the development of the celebrated Swatch, 10 years later, did the Swiss industry recover. This precedent ought to serve as a lesson.

Consider the lock industry, which is doomed in its traditional form. The future does not rest on purely mechanical devices to secure a door. The training of locksmiths can no longer be limited to a knowledge of bolts and catches, but must embrace microelectronics as well as new technologies, such as magnetic sensors. Some lock manufacturers have sensed the shift of the wind. And it is not done shifting! Already other devices are poised to supplant cipher cards: devices based on recognizing hands, voices, retinas, and so forth. This evolution demands long-term research.

Maintaining strong relationships between scientists and industry, through a sustained and genuine dialogue (including sabbatical leaves for university researchers), is in my view one of the necessary conditions for survival as a developed society.

A Tennis Game

Industry provides suggestions, penetrating questions. It is like a tennis game. In my field, the spark frequently comes from the industrial worker who succeeds in manufacturing something totally new. Recently, for example, cows have been made to ingest encapsulated amino acids. The proteins are, therefore, not broken down by the trypsin in the bovine stomachs, but are instead passed to the intestine, where they can be directly assimilated. This type of product may well revolutionize the way cattle are fed. It is no ordinary feat to develop a coating, an encapsulation, that resists the multiple stomachs of cows while permitting subsequent diffusion and absorption of the substance at the right time. This is a perfect example of an industrial innovation that required scientific thinking.

Exchanges between industry and universities do not require large investments; nor are they one-way. The benefits are mutual. But while our corporations need to promote dialogue between fundamental scientists and their industrial counterparts, openness is not always possible. It is not easy to conduct a discussion that impinges on the area of trade secrets. Sometimes, an extraordinary idea is revealed very late in the game because its premature disclosure would have placed an enterprise in jeopardy. Secrecy is particularly justified in the case of the small to medium enterprises, which are more vulnerable, in case of leaks, than large corporations.

Much remains to be done for smaller businesses—in particular, in the area of the agricultural and food industries. These industries constitute one of our greatest resources, one of the sectors with the strongest growth potential in our country. Agro-food firms of small to medium size are legion, but right now we have difficulties approaching them. In this context, I cannot resist mentioning a particular success, modest as it may be, of which we are particularly proud at our Institute. We have a group, engaged in the field of acoustics, that has been granted a patent for the "probing of camembert cheese." Instead of sticking one's thumb into the paste to assess its consistency, as is done traditionally, it is possible to use a noninvasive acoustical technique known as speckle holography. This sensitive method permits the probing not just of camembert but of most cheeses available on supermarket shelves.

Generally speaking, there is a dearth of bridges to reach the small to medium businesses. There exist research organizations available to work on contract, like Batelle in Switzerland, or Bertin in France, to which, in principle, a small firm could go and commission a study on a more competitive process. In practice, however, these research organizations deal almost exclusively with larger clients. France's National Center for Scientific Research has created information centers to assist small to medium businesses, but more should be done. Perhaps it should be stipulated that each young researcher on the public payroll must spend one day a month in a medium-sized enterprise.

CHAPTER 2

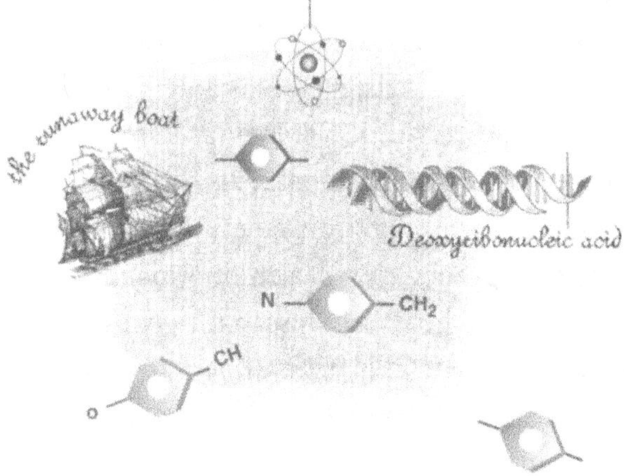

Discovery

Edison and Feynman

Thomas Edison, the prolific inventor, was fond of saying that science requires "5 percent inspiration and 95 percent perspiration." The formula still applies today. It takes a great deal of endurance and persistence to produce something new, even under truly favorable conditions. Recently, our team at the Collège de France was investigating the way a molten plastic flows inside a tube. It is an important problem—for instance, in the process of extrusion of plastic bottles—one that involves poorly understood phenomena. Studying the problem requires measuring speeds in very close proximity to a solid surface (within 500 Å, or 2 millionths of an inch) to determine whether the polymer glides on or sticks to the surface of the tube. This can be done with sophisticated optics. An experiment conducted by applied researchers using optics techniques took 5 years to set up, but it produced beautiful results. It demonstrated that, on carefully prepared surfaces, the polymer sticks under a low-speed regime and glides under a higher-speed regime. By happenstance, the theorists in our group found, almost at the same time, an explanation for the transition between two regimes that agreed perfectly with the results of the experiments.

But this was an unusual stroke of luck. Most of the time, there is great confusion on both sides. One has to start over, to search, to grope, guided here by a theoretical principle, there by an experimental innovation. For example, new microscopy techniques (for example, scanning-probe microscopes) are available today which enable us to examine things on a scale not of a micron, but a thousand times smaller. These microscopic probes call for new concepts and open up new areas for thinking.

In general, events are set in motion by a provocative question (a completely mysterious phenomenon that piques the researcher's curiosity, for example) and by a patiently nurtured expertise. These conditions are necessary, but not always sufficient: lines of thought that appear solid at first blush collapse miserably after a crucial experiment, or because of a change in scale. One must constantly correct the aim or even change ideas outright.

What role does theory play in discovery? Does it provide strict laws? These are questions that I hear often. One succinct answer was given by a great American scientist to whom my generation owes a great deal: The elementary particle theorist Richard Feynman, had a wonderful definition of theory: "Theory is the best guess." This definition is full of humility, and it also happens to be right.

What role does luck play? It is undeniable that every piece of research involves some degree of gambling. But luck is only one factor among many. If you see a particular team repeatedly hit paydirt, it is mostly because it does not shy away from hard work.

Memories of a Trip

To jump in or not? The decision to enter a particular field carries no guarantee of success. But even before one has to make this kind of decision, one faces other crossroads, and the road one takes will depend on a subtle mix of cultural formation, personal choices, and the vagaries of life.

I am acutely aware that I belong to a generation fortunate to blossom under favorable conditions. Too young to be caught in the tragedies of the war, I entered research at precisely the right time (around 1950) to benefit from exceptionally fortunate circumstances: France was in full reconstruction and was in the process of making up its scientific lag in spectacular fashion. I was guided at the very beginning of my career by researchers who had been trained abroad, and come back home to teach us, the spoiled generation, everything that was new: in my own case, teachers like Albert Messiah, Anatole Abragam, Claude Bloch, and Jacques Friedel, people who, essentially, founded the French school of modern physics.

My counterparts just three or four years older than I did not all enjoy the same opportunity. They found themselves in a period when France had neither the teachers nor the books to educate them. I think, for example, of quantum mechanics, a science based on waves, indispensable today to our understanding of atoms and nuclei. My seniors were often exposed to quantum mechanics toward the tail end of their studies, too late to truly master the subject. For some, it proved to be a life-long handicap, not unlike the handicap of an individual who takes up the cello at 16 years of age: the virtuosity and agility of the fingers will never match those of someone who started at a much earlier age. The building of a knowledge base stretches over a period of one or two years.

In short, I was extremely lucky. And that luck included my first job. More often than one cares to admit, a job is chosen on the basis of sentimental, sometimes completely irrational, criteria. That is very much how I decided to join the Atomic Energy Commission. Jacques Yvon, one of the managers at the Saclay Atomic Research Center, sent a chauffeured limousine to take me to my first interview. I felt like a king! The second bait was equally intoxicating. Upon arriving at Saclay, I saw these gigantic buildings: particle accelerators. In complete awe, I signed on the spot to work there. I could have made a colossal blunder with that kind of motive. God knows that the modern world has its share of huge buildings that are totally useless! But it was my lucky day, and things turned out beautifully. The atomic energy establishment had truly creative people; among them was my boss, André Herpin, who was to become a dear friend. Under his guidance, I moved from quantum physics to its natural offspring, solid-state physics. Next, I worked on magnetic materials. Magnetism was at the time a science in full development, largely under the leadership of Louis Néel in Grenoble.

Research reactors, like those at Saclay, provide, by virtue of the neutrons they produce, a new method of study of magnetic materials. Under the guidance of Bernard Jacrot, I became interested in the technique. Later on, after a delightful year working at Berkeley, California, under another boss, Charles Kittel, who also became a friend, I spent 24 months in military service. It was upon my return that, as a young lecturer at Orsay, I had the naive idea of directing an experimental group working on superconducting metals, materials which conduct electricity without losses at very low temperatures. The general principles of this extraordinary phenomenon had just been worked out, after 50 years of groping in the dark. The superconductivity group at Orsay experienced a difficult beginning. But thanks to the help of a few outstanding colleagues, like P. Perio, A. Guinier, and J. Friedel, it became quite successful (1962–68).

In 1968, at the CEA, we became interested in an entirely different topic: liquid crystals, which had been a very active field in France around 1920, thanks to George Friedel. The Russians had picked up the baton in the 1930s, but their fetishist trust in things theoretical caused them to give up the field, which they considered too "chemical" and "dirty." On the eve of the events of 1968, an article published in a Soviet journal made us aware of the history of the subject. The article contained a comprehensive review of the knowledge acquired by the Russians in the domain of liquid crystals. For us, it was a tremendous revelation. There were enormous gaps, fascinating aspects that the Russians had missed—a veritable gold mine! Within a few months, every member of every group associated with ours understood that this opportunity had to be exploited. Six groups (crystallography, chemistry, materials defects, optics, nuclear resonance, theory) switched in lockstep, with their entire complement of equipment, to the field of liquid crystals. There was no need to create new laboratories. With minimal financial investment and in a very short time, we had achieved a respectable international position. The times were propitious; minds were open and favorably disposed toward collaboration.

A few years later, we switched once again, retiring from the field when it was no longer possible to conduct fundamental experiments within our limited means. We turned our attention to polymers, which in the early 1970s posed intriguing questions. As fate would have it, neutrons proved once again to be an excellent tool for studying polymers. This was quickly understood and exploited by my friend Gérard Janninck at Saclay and, later, by a number of other active groups, notably that of Henri Benoit in Strasbourg.

Later, around 1978, we switched to the topic of colloids—ultra finely divided matter—and, more generally, to the problems of interfaces and, eventually, to the mechanisms of adhesion, in other words, what makes a glue effective.

This, in a nutshell, was the path we followed. At each stage, we committed ourselves because we had at our disposal conceptual tools and experimental techniques which we felt had the potential to be productive.

The media have a tendency to portray us (myself or other researchers of the same ilk) as flitting butterflies, superficial dilettantes, jumping from subject to subject. I should be so lucky as to have this gift of universality. The reality is that, every time one shifts to a new field, one has to catch up with the rest of the class, to learn all over again from scratch! When we got started with polymers, we were staring at four long years before we could contribute our first useful idea. The macromolecules laboratory in Strasbourg, founded by Charles Sadron, was enormously helpful to us in the beginning, when our ignorance of the subject would cause us to make incredibly absurd statements. Thanks to the patience and intelligence of that team, led by Henri Benoit, we got educated. Eventually, we were able to propose a few ideas which were novel precisely because we came from a different world. But first, we had to learn everything our predecessors had done.

Half-Way Across the Ford

At the moment, my main interest revolves around the problems of adhesion. I have said little about this topic so far, but it illustrates the dilemma faced by the researcher choosing a subject and trying to contribute to it. Our trip down this road, too, has stretched to over four years.

The history of adhesives is a long one indeed. Man has known how to glue since early antiquity. The Assyrians and the Romans used isinglass, also known as fish glue. The Phoenicians discovered products that were both adhesive and waterproof, a technical innovation that greatly contributed to their maritime supremacy. For centuries, furniture craftsmen preserved the art of gluing, and then it suddenly exploded at the turn of the twentieth century with two major developments:

- The discovery of synthetic polymers (today, all adhesives are polymers).
- The birth of the aeronautics industry (until the 1930s, most airplanes had wooden frames, which had to be assembled with glue).

Progress has been spectacular. Today, we have access to a wide range of adhesives that make it possible, for instance, to glue rubber to metal, a feat which was virtually impossible in my younger years. These successes have been remarkable from a practical standpoint and adhesion has created an important industry. But to teach this science remains rather frustrating because one cannot give the students much more than a compendium of recipes.

When we, soft matter physicists, were considering working on the problem of adhesion, did we make a wise bet? It is too early to tell, as we are only half-way across the ford. We are trying hard to extract a few simple principles, a few clarifying ideas. We are just beginning to understand the adhesion of pliable materials, or even pliable materials on hard ones: the concepts are firming up. But when it comes to the subclass of epoxies, where the polymer is in a glassy state resembling window glass, we are still in total darkness.

Our situation as researchers can be compared to that of a Livingstone or a Brazza, of an explorer of the African continent in the last century. We are struggling to find the source of the Nile by sailing up various tributaries. The process has nothing to do with an esthetically pleasing exercise conducted in the style of a flawless demonstration.

In France, a small number of theorists and a few experimental groups are involved in adhesion research. We have at our disposal an excellent laboratory in Mulhouse, with a good mix of physics and chemistry. But there is not enough interaction with engineers. Our most brilliant engineers are poorly trained in chemistry and in the science of polymers. As it happens, adhesion involves many sciences. I hope that a group can be created in the Paris region uniting the talents of chemists, physicists, and engineers to focus on the subject.

The Nanonewton

My interest in the mechanisms of adhesion prompted some rather foolish comments in some of the media at the time the Nobel Prize was announced. One journalist gave me credit for inventing glues, which obviously did not have to wait for me to flourish. Another, during a televised interview, confused colloids, one form of ultra finely divided matter, with the science of glues.[1] Others linked the Prize with work on epoxies, which is precisely the area which I least understand! In another vein, a few even compared me with Newton! If I have anything in common with that great man, it is on the level of a nanonewton, a unit of force that is about equal to what it takes to break a chemical bond, and which is measured in certain experiments in adhesion.

Having said that, I am not adverse to using Newton as a model, but for reasons which should evoke in researchers feelings of awe and humility. Isaac Newton was formidable, ferocious, passionate, scientifically dazzling from an early age. He had a spectacular résumé: at age 13, he was building in the English countryside machines that fascinated

[1] Translator's note: The confusion is more transparent in French, where the word *colle* ("glue") sounds much like *colloid.*

the neighborhood. At 18, he invented the reflecting telescope. Around age 20, he understood optics in terms of the wave nature of light. Between 20 and 25, he explained gravitation and the motion of the planets. As a bonus, and for good measure, he simultaneously made a fundamental contribution to a branch of modern mathematics: he invented differential and integral calculus. In short, a flawless track record. Suddenly, around 27 or 28 years of age, he switched subjects and wandered off toward alchemy, to which he devoted 10 completely wasted years.

This lost gamble constitutes one of the most frustrating episodes of the history of science.[2] Newton had within his reach the tools which, one century later, would enable people like Lavoisier to lay the foundations of chemistry. Instead, Newton, orthodox Protestant, strictly respectful of the Book and of God's Word, believed that he had found in the formalism of alchemy—through its structure, its numbers, and its classification schemes based on geometrical figures—the key to knowledge. In that frame of mind, he lost himself in an esoteric maze worthy of the *Romance of the Rose*, of medieval fame.

In spite of it all, he later on derived some dividend from that experience when he became England's Master of the Mint under the reign of William of Orange. Having gained some skills in chemistry, he had the wherewithal to ferret out currency swindlers bent on stealing gold or silver by cheating on the composition of alloys.

Blunders

Scientists deeply involved in searching for answers are apt to make foolish pronouncements. Lord Rayleigh, one of the greatest engineers of the last century, wrote shamelessly a few years before Clément Adler's first flight in the *Éole* that to attempt to fly a piece of machinery heavier than air was a sheer waste of time! In 1933, Ernest Rutherford (1871–1937), physics Nobel laureate in 1908, described the prospect of exploiting nuclear energy as a "fairy tale." Less than 10 years later, in 1942, the Italian physicist Enrico Fermi (1901–54) built in Chicago the

[2] See L. Verlet's *La Malle de Newton* (*Newton's Trunk*) (Paris: Gallimard, 1993).

first atomic reactor. But no physicist could possibly have foreseen such developments at the beginning of the 1930s.

A certain segment of the literary world grants science anticipatory powers quite disconnected from reality. In *Majorana's Disappearance*, the great Sicilian writer Leonardo Sciascia engages his readers in a story made to appear real, but which is actually a theory invented by the author. Ettore Majorana was a young researcher, a colleague of Enrico Fermi's. In 1936, he went through a period of deep depression. After isolating himself for six months in a room, he sailed from Naples to Palermo and disappeared without a trace. Sciascia concludes from this that the man was haunted by the terrifying vision of weapons which could be produced on the basis of the physics of nuclei. The hypothesis is actually quite unrealistic. I queried a number of Italian physicists who were working during that period (1932–36). They unanimously confirmed to me that no one at that time had any concept whatsoever of a chain reaction using uranium. Besides, Majorana was actually a theoretician working on abstract problems in quantum mechanics. He was in no way familiar with the practical aspects of nuclear physics. But writers sometimes prove to be more adept at theory than the theorists themselves.

Scientists are not wizards. They do not divine the future, even though they are busy building it. I myself remember at the end of graduate school thinking of optics as a dead science. I was convinced that everything was understood that could be understood, and all that was left amounted to nit-picking. I was later forced to eat my words after certain subsequent advances, like the invention of the laser. The laser was followed by other extraordinary discoveries, such as nonlinear optics. The science of optics is like a phoenix perpetually rising from its ashes.

There was a similar lack of judgment on my part about superconductivity, a field with which I became well acquainted in my early career. If I had been asked to assess the future of that science a decade ago, I would have given some embarrassing answers. Everything seemed to me to have been worked out. Theory had established that, with usual metals, superconductivity could exist only at temperatures far too low for industrial exploitation. There were, of course, a few interesting but limited applications, like the production of high magnetic fields, as implemented so spectacularly in medical imaging systems using nuclear magnetic resonance.

As a result, I was an advocate of a small-scale, stand-by effort in that field, and recommended that the bulk of the troops convert to other areas. As it happens, the science of superconductivity is now in the midst of an explosive rebirth. A few months after the discovery (in 1986) by Alex Müller and his co-workers at IBM's Zürich laboratory of new compounds, based on oxides, that remain superconducting at temperatures far higher than conventional materials, thousands of researchers climbed onto the bandwagon. It is still an area of highly active research. Why do these oxides enjoy such unexpected properties? After ten years of efforts, the answer is still not clear. At least three schools of thought compete. The climate has grown tense, to the point that separate conferences must be held to avoid coming to blows. I personally prefer to rummage about with a small team in a discipline with less visibility, at least for now.

Know When to Stop, Know When to Switch

At the moment, *nuclear* physics, the science of atomic nuclei, has in my opinion reached the end of the line. It made enormous strides in France, notably under the auspices of the Atomic Energy Commission (CEA) during the 1960s, and spawned our stock of nuclear power plants. This branch of physics requires costly support and has developed, incidentally, a powerful lobby. But it seems to me to be technologically drained. At the risk of offending some of my colleagues, I am tempted to shout: "Enough!" But not indiscriminately. That would be as absurd as to try to stop a high-speed train. A wiser strategy would be to shunt the train onto another track, more novel, and more useful to society.

To reorient a science as ponderous as nuclear physics is no easy task. The problem is to divert people in directions that are both attractive and useful, to convert, for example, a number of nuclear scientists into molecular collision physicists, or into atmospheric scientists. My own experience is that it takes three years of hard work to change. For a theorist, three years is still in the realm of the feasible. For an experimenter, it is much harder. When, in the mid-1960s, I left the physics of very low temperatures and of superconductivity to go into that of liquid crystals, a handful of co-workers of mine, a few experimenters with a great deal of fortitude followed me. To change research fields is as traumatic as moving to another country. An experimenter who switches fields loses a language, all the books that were familiar to him, but most importantly, all his tools and recipes. He must effectively go back to school. He is a traveler without suitcases. He must cut himself off from his network of friends across the world. This system of contacts is very important. It keeps you constantly informed. If, for instance, a new product has just been synthesized by a chemist, he often makes a sample available to you. All of this disappears overnight.

On the other hand, it is not acceptable to coast along in a structure that has grown too rigid, with well-entrenched habits, patently vegetating at the taxpayer's expense. The inertia of the scientific elite was endemic in the republics of the former Soviet Union. Toward the end of the 1980s, I tried to warn my Russian physicist colleagues: "Watch out, or you will find yourself in a situation of famine. You will no longer be paid unless you quickly justify your existence to your fellow countrymen, who will themselves be struggling in terrible misery." They did not believe me. Soviet physicists had been groomed in a system which placed them in a greenhouse, shielded from public inquiries. They were bureaucrats, whom nothing could threaten. Lately, they have begun to discover reality and are scrambling (a bit late) to turn their attention toward a more useful science.

To convince people of the necessity to move, one has to preach, and preach tirelessly. I am glad that I succeeded in one particular case, even though it was only on a very small scale involving just about twenty researchers at the Collège de France. At the time, all were involved in polymer problems. One day, while talking with two other theorists, I commented that the field had become quite mundane, while there were fascinating questions in the area of the dynamics of wetting. I started discussing the issue throughout the laboratory, and a miracle occurred: within a year and a half, three groups had switched. They had to dare. But these particular researchers had unusually open minds. Our success was also due to the young age of our staff. The turnover was rapid. People were not overly tied to their old disciplines.

The passing years also have a tendency to prompt you to change subject. In the field of superconductivity, my group had created toward the end of the 1960s a generation of researchers who had become more productive than I. The time had come for them to take the initiative, to take their own lumps, under their own responsibility, instead of relying on me as a sort of grand-uncle, dispensing advice.

Explosions and Bengal Lights

Like painting, which passed within a century from Ingres's classicism to impressionism, and later to different forms of abstract art, science alternates between periods of upheaval and periods of strengthening. There are eruptive phases and others, no less extraordinary, of validation and consolidation. It is important to perceive the times correctly.

The nineteenth century emphasized chemistry; the first half of the twentieth, physics.[3] Right now, molecular biology is at the forefront of scientific breakthroughs. Tomorrow may be the turn of the physiology of the brain. The world will undoubtedly have other revolutions comparable to those caused by the works of Galileo, of Newton, and, in atomic physics, of Schrödinger and Dirac. But the public forgets that these explosions also generate more durable Bengal lights. When Pierre Simon de Laplace, Thomas Young's rival, began crafting his *Treatise on Celestial Mechanics*, he relied on Newton's ideas, whose influence was to remain preeminent for the next two centuries.

No researcher can foresee all the potential consequences of his ideas. They might germinate along the way, through a long maturation process. At the present time, an effort is taking place to deepen our knowledge in highly specialized, quite fundamental areas, like elementary particle physics. The general public is not always aware of this because discoveries do not manifest themselves in the form of visible applications to daily life.

[3] In my opinion, the finest contemporary book dealing with the development of physics is without a doubt Freeman Dyson's *Disturbing the Universe*.

In astrophysics, revolutions come in quick succession. I am not in any way a specialist of cosmology, but discoveries in that field constitute for me a constant delight. I remember, as a young professor, teaching my students a purely theoretical idea, advanced by J. Robert Oppenheimer and others, according to which a supernova, which is a collapsing star, should end up as a sort of extremely dense ball made of neutrons. I would describe to them this strange object, never before observed, with unusual properties: a neutron star, as heavy as the sun, but only 10 kilometers in diameter, on the surface of which the smallest seismic wave, only a few millimeters in amplitude, would have consequences visible from Earth. Truly quite a bizarre world. As fate would have it, I was forced to update my lecture the next time I was teaching the course. I no longer had to talk about an invisible object: it had become quite visible. Rockets equipped with X-ray detectors had just returned data demonstrating the existence of anomalous X-ray sources. And the only way to explain them was to invoke neutron stars.

There will always be new frontiers to cross. Some of them were identified very early on, like those imposed by the notion of time. Telescopes equipped with increasingly sensitive instruments peer ever farther into space. And the greater the distance, the farther back in time one travels. Conversely, particle accelerators explore ever higher energies, which affords the probing of ever smaller regions of space. These types of frontiers will never cease to exist.

Other frontiers are closer to us, and perhaps more accessible as well: the workings of the brain are on the verge of becoming the hottest topic. Our younger generation will in all likelihood become involved in this type of research (and deal with all the dangers that it implies).

Mysteries are bound to be unlocked in the field of socioeconomics as well. This science is still in its infancy. But here again, everything will depend on the cultural and instrumental context alluded to at the beginning of this chapter. Science can bear fruits only to the extent that it has matured; to force the growth of a science without the appropriate technical and conceptual tools is to condemn oneself to failure.

Finally, it is not always necessary to look in obscure corners to come up with intriguing questions. Plenty of them, sometimes unassuming ones, exist in plain view. We come in daily contact with familiar objects, even gadgets designed to entertain children, which are not always understood by physics. I think, for example, of a popular toy sold in supermarkets which consists of a small piece of elongated plastic resembling the hull of a slightly twisted canoe. If you place it on a table and give it a slight vertical nudge to cause the bow to drop, it starts spinning sideways, always in the same direction. If you try to force it to turn in the other direction by nudging the bow laterally, it complies for a few moments, only to stop and start again in its naturally preferred direction. Why is that? A few calculations have been attempted, but we frankly lack a simple explanation.

CHAPTER 3

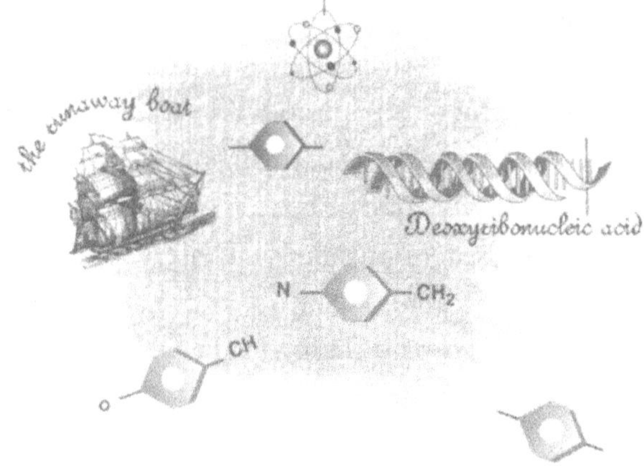

A Positive Science

The Duty to Inform, Not the Power to Decide

The public image of scientists in our Western world is often inaccurate. Either the public accord them the power of prophets, or it perceives them as sorcerer's apprentices. In either case, the stereotype is distorted.

The primary public responsibility of a scientist is to inform his fellow countrymen without delay when he sees an emerging application or concept which is likely to pose a problem. It might be a pill, a weapon, a technical process. But scientists are not entitled to decide what is good or what is bad for their contemporaries. I am against a government of technocrats or of experts in general. In a democracy, the responsibility for the choices facing society belongs with the citizens and their elected officials. The scientist's responsibility, beyond his work, is merely to warn that some decisions have to be made.

In 1941 a number of Western researchers understood the military applications of the process of fission [8] based on the chain reaction sustained by uranium. It became conceivable by then to build an explosive device thousands or even millions of times more powerful than anything devised up to that point. They met with President F.D. Roosevelt to inform him. As the individual responsible for the executive branch, it was his duty to decide whether the United States should pursue this weapon. Roosevelt opted—justifiably, it would seem—in favor of the atomic bomb. It was impossible to passively twiddle one's thumbs, given the scientific and technical power of Nazi Germany: it, too, could well start developing that weapon.[1]

This problem has changed little during mankind's entire history. From the poisoned darts used by Amazonian tribes to the war implements built by Archimedes for Dionysius of Syracuse to the H-bomb, scientists and technicians have always offered to their tribe, to their nation, a possibility, but not a necessity, of making use of new technologies for good or for evil. I cannot deny the catastrophes; there are plenty of them, and of all types. But these have more to do with man's ignorance, greed, and vanity than with the progress of knowledge and technology.

When the American biologist G.G. Pincus (1903–67) developed in 1955 the first birth control pill blocking ovulation, he did not step in the shoes of the legislator who had to decide whether to legalize its sale. Society may not abdicate its responsibility in favor of researchers under the pretext that it might make an evil use of their inventions.

Research provides an idea, not the method of its social usage. That decision rests with the collective body. When a medical association decides to oppose abortion, as apparently happened in Poland, it is guilty of abuse of power. It is up to the citizenry to resolve this kind of public policy matter. The scientist is not a demiurge; he is no more dependable than anyone else when it comes to moral, political, or judicial issues.

I am also suspicious of the attitude of some ecologists who would like to ban certain types of research. There will always be someone ready

---

[1] German scientists did not build the bomb, not because of ethical concerns, but primarily because they had overestimated the amount of plutonium required.

to start a project, with or without official permission. My own sense is that the risks would be far less if the activity were sanctioned but controlled. Without oversight, a different breed of people might be performing the work, driven by dubious motives—acting as mercenaries, operating in secrecy, disregarding precautions, and making the whole venture that much more dangerous.

In research as in other human endeavors, the risk of dictatorship is always present. I am fearful of some examples that are rising over the horizon: Islamic fundamentalism, for instance, or the so-called "politically correct" movement in the United States. I would summarize my view by making two points:

- We need regular consultations between the scientific community and elected lawmakers. There is a need for collegial contacts for the purpose of reflecting on long-term outlooks and of proposing solutions to public officials—in short, a pragmatic and balanced approach to problems. France is not without stature in this domain. The national Ethics Committee, of which I can talk with ease since I am not a member, seems to me to be a useful and intelligently structured forum.
- We must improve the ability of public opinion to make the correct choices. At present, too many hidden pressures, and too much irrelevant data, serve only to confuse the collective decision-making process. The level of scientific education of our youth must be raised. They have to be lucid in the face of the media.

One Danger Can Hide Another

The debate surrounding genetic engineering currently occupies center stage. In my judgment, the short-term danger lies not so much in these sophisticated techniques, which are difficult to acquire, as in the proliferation of bacteriological weapons. Their production is within the reach of a great many nations. Even more threatening, and largely ignored by public opinion, are the problems that are sure to come up in connection with brain research. Suppose that a drug is found that stimulates a specific cognitive or motor function, but at the same time possesses negative side effects? What should be done then? Should the product be reserved for a few? Should it be made available to all? Should its use be banned? Our emerging knowledge of the physiology of the brain is likely to confront society with difficult choices. Painful ethical dilemmas await us in this area.

Physics and Metaphysics

In my student days, I was sometimes irritated by philosophers. I found myself among literary theorists, devoid of any experience of practical life. They would expatiate on any and every subject with extraordinary self-assurance (and a great elegance of speech).

Obviously, philosophy must have a voice in the choices presented to society by scientific progress. It is impossible to live without exploring beyond what is accepted as the norm. Unfortunately, the example of the Age of Enlightment illustrates how theoretical philosophers, like J.J. Rousseau, created a climate of thought that culminated in a bloody and senseless revolution. We can only hope that history will not repeat itself in the West under the influence of the "politically correct" movement.

My generation was traumatized by the issue of political involvement. In my age group, at the École Normale Supérieure,[2] many of my classmates, often the brightest and most motivated, were hard-core Marxists. One of the teachers on the faculty of that school was a Marxist philosopher by the name of Louis Althusser, who had a reputation as an infallible oracle. These highly respectable people were headed toward an impasse. Had they gained power, where would they have taken us? Coming in contact in my youth with one such doctrinaire attitude has made me distrustful of political doctrines in general.

---

[2] Translator's note: The École Normale Supérieure is one of France's premier institutes of higher education.

A high school senior once asked me to comment on two questions of philosophy posed during final exams: "Does the physicist deal with reality?" and "Does matter exist?" These are obscure questions, to say the least. I am shocked that our high school graduates, who know so little about life, are selected on the basis of their ability to concoct a theoretical discourse on such esoteric topics. I simply gave a naive answer to the first question.

What we call reality seems to me to be the baggage of experiences and issues which we faced during the course of our youth. More generally, many abstract definitions actually derive from a *learning process.* Such is the case, for instance, in our small scientific world with the definition of "physical intuition." When I declare a proposition to be physically intuitive I imply that it conforms with what I have learned in the laboratory or in physics textbooks between the ages of 18 and 21.

The time of the great battles between religions and the sciences seems, in my view, to be over in the West. The Churches have admitted that they have nothing to gain by trying to justify their theological postulates with scientific arguments or trying to influence the progress of science. This was not always the case. The most famous example remains the condemnation, in 1633, by the Holy See of Galileo's thesis concerning the motion of the Earth. Today, despite isolated incidents, it is fair to say that, in the physical sciences at least, the roads have diverged and we are no longer on a collision course.

On the question of metaphysics, researchers are just as divided as their lay contemporaries. Some have religious beliefs, while others are resolutely agnostic. Given the present state of our knowledge, nothing permits us to exclude a Grand Project of the universe. Nor is there any persuasive evidence forcing us to include such a point of view in our work. Nevertheless, a few cannot resist the temptation. Some argue that the Big Bang is evidence of a divine intervention in the creation of the universe (the Big Bang model postulates that the universe was born in a gigantic explosion). In fact, we do not even know if the Big Bang is a unique event giving birth to a universe that expands forever, or if it is a recurrent event in a universe that expands up to a certain point only to contract again and come back to zero.

Others debate the origin of life. During the 1950s, a plausible scenario for the appearance of life enjoyed fairly wide credibility. Simple organic molecules were believed to have been simmering in a great ocean stew. Extreme conditions of ultraviolet radiation and electrical discharges would have fostered the formation of increasingly complex molecules, ultimately leading to the emergence of life. Today, the trend has shifted, it seems to me, toward a more cautious posture. The emergence of life remains, for several reasons, a poorly understood phenomenon.

Amazingly enough, the genetic code is identical for all living beings. This is peculiar since evolution subsequently did not preclude the coexistence of a variety of systems, such as vegetal and animal, vertebrates and invertebrates, and so forth. How can we explain the fact that natural selection did not favor the simultaneous development of several forms of the genetic code from a common basis? This mystery is quite puzzling.

The biologist Francis Crick, one of the codiscoverers of the double helix—the molecular structure of DNA—is apparently convinced that Earth was seeded, in some fashion, with a particular code bequeathed to it. What might the source of this seeding process be? Crick talks in terms of an extraterrestrial origin, the legacy of the death throes of another world. We have no experimental means to ascertain the answer. We can only wait, and, above all, refrain from hasty theological conclusions.

The phenomenon of chirality is another one shrouded in uncertainty. The word *chirality* comes from the Greek *kheir*, meaning "hand"; it refers to the fact that a molecule can differ from its mirror image the way a left hand differs from a right hand. We are all made of amino acids with a particular chirality. If one synthesizes amino acids in the laboratory, the result is 50 percent of the left-handed variety, and 50 percent of the right-handed variety. But living matter turns out to be 100 percent left-handed. Why? If one accepts Francis Crick's conjecture, it is possible that we were "seeded" exclusively with left-handed molecules. But what if we were not? A possibility, which I discussed some time ago, would be that right-handed (R) organisms, through some hypothetical competition mechanism, would develop less readily than their left-handed (L) counterparts when in the presence of a majority of L. Similarly, L would be less apt to grow in the presence of a majority of R. On the basis of this postulate, one would expect the development of a spotted faunal map, with L zones and R zones, separated by boundaries which would be the seat of L/R competition. It is then possible that, through simple statistical fluctuations, a particular L zone might expand and occupy the entire Earth. Yet, with exactly the same scenario, we could just as well have ended up in a R form. At any rate, this remains a purely abstract model, and we are a long way from testing this hypothesis with experiments.

In summary, one can argue that the existence of an identical genetic code in all living species and the chirality of the living world partly dispel the myth that science has made any headway on the issue of the origin of life. That said, these observations, which for the time being leave both the biologist and the physicist perplexed, have, in my own mind, no real metaphysical impact.

Measure Before You Judge

When faced with phenomena in everyday life that are unsatisfactorily explained, or not explained at all, the best strategy is to refrain from espousing any a priori doctrine. If something a bit strange presents itself, the proper course of action is to try to conduct precise measurements. A good example is that of *dowsers*, who were of considerable interest to Yves Rocard (Michel Rocard's father), himself a great scientist, engineer, and industrialist from the 1930s on, a pragmatic and omniscient mind, and who was one of my teachers at the École Normale Supérieure.

For thousands of years, dowsers have been called in to help before digging a well in one's garden. Nobody really knows how, or even if, dowsers are capable of detecting the presence of water. Yves Rocard himself failed to propose a convincing mechanism. Nevertheless, one could conceive of a systematic experimental program to find the answer. For example, a network of water pipes could be built underground, and dowsers allowed to roam about, a few meters above, in order to monitor their signals to see if they might be able to reconstruct the architecture of the network. But this experiment requires a lot of precautions: it is possible, for example, that the pipe material might cancel or contaminate the effect. These days, no agency would dare sponsor such a research project.

CHAPTER 4

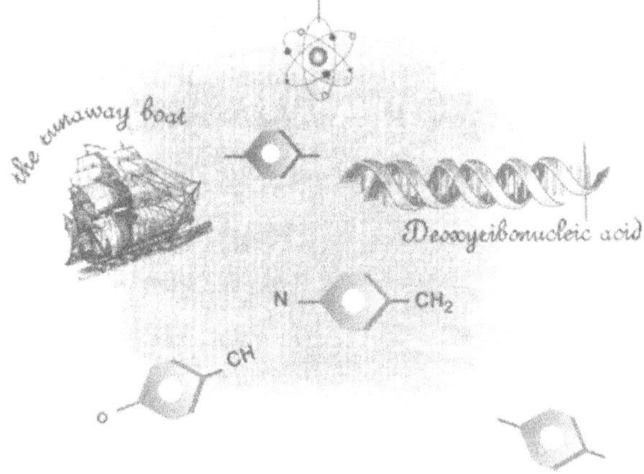

The Environment

Making Science Heard

Successfully completing an experiment or constructing a theory that describes reality provides a happiness that one would like to share. But the cultural barrier is sometimes insurmountable. How can one make transparent to nonspecialists the esthetic appeal of a science like elementary particle physics? Exploring nature on the atomic and subatomic scales requires years of scientific training to master the language and to begin the process of understanding. Certain joys and certain conquests will have to wait a long time before they are recognized beyond the circle of a tiny community of experts. This obstacle can be painful to many researchers, myself included.

J.P. Sartre once wrote something like: "Three bars of music, and one is hooked. . . ." It is quite true. The musician enjoys the wonderful privilege of being able to share with his audience the beauty of a melodic line in just a few seconds. By contrast, a scientist can convey what he perceives as beautiful and important only at the cost of tedious explanations—in the best of cases with lots of drawings and diagrams. In subjects more complex than soft matter, it is even more difficult.

I have come across young researchers who have done beautiful experiments and who would dearly love to share them with the public. But I hesitate to send them off too often in that direction, because one cannot do everything at once. Following the Nobel Prize, I tried to address this communication problem by touring high schools. But, as I have already confessed, I could not sustain the pace longer than 18 months. The price exacted by fame and the media is to have the spotlight trained on a single individual. I know at least 20 physicists who could give better presentations than I to an audience of high school students. Unfortunately, public relations demand that one call upon the one researcher who recently stepped on the podium. To the schools that I was unable to visit I submitted a list of possible alternatives, in the hope that our prospective hosts would prove more mature than the media. It is time to establish between our profession and the world of secondary education bridges that are not at the mercy of current events.

Many disciplines lend themselves to lively presentations. Geophysicists can tell amazing stories about the life of our planet, about the extinction of dinosaurs. Hydrodynamics is another case in point. Many students are afraid of the subject because they expect to be inundated with calculations. In fact, hydrodynamicists are making fascinating discoveries that can be understood by the man on the street.

In an area as little known as the mechanics of powders, there is much to tell. During a television interview, I demonstrated an experiment involving powders (which, by the way, did not work very well!). The experiment consists in causing a pan containing salt or sand to vibrate vertically by placing it above a horizontal loudspeaker. When the amplitude of the vibrations reaches a certain threshold, the shape of the powder layer changes and takes on the appearance of a volcano: a miniature Fujiyama. Why is that? Nobody knows exactly. This phenomenon was discovered by two Frenchmen, Pierre Evesques and Jean Rajchenbach. After my televised demonstration, I received a flood of enthusiastic letters purporting to give elementary explanations of the effect. In fact, the mechanics of powders is far from being elucidated. Even a gadget as plain and common as an hourglass is still not satisfactorily explained. Yet there is a need to understand. Powders are used in steel mills to drive the liquid metal in a mold; the same empirical approach is used in grain silos. By the same token, the mechanics of powders are applicable to the the mechanics of sedimentation in river deltas, about which very little is known.

Another area where science can say something about daily life is the dynamics of washing and rinsing. Many people would like to understand the objects they use routinely a little better, and it is possible to provide them with simple explanations.

I had the opportunity to discuss these issues with 200 primary school teachers from the nineteenth district of Paris. It was an emotional occasion for me, because the profession of school teacher is one of the few, besides mine, that I would have liked to go into. (In my youth, I got to know some marvelous school teachers in remote mountain areas. They had to teach, under difficult conditions, children of various levels and ages, all gathered in a single room, and managed to steer them successfully toward a secondary education. I have an infinite amount of respect for this profession.) The teachers had invited me to discuss with them the impact of science at the primary education level. I admired this group all the more because primary schools in the nineteenth district handle children whose families often experience ethnic integration problems, leading to difficult situations in school. Still, these teachers had taken the time to reflect on possible pedagogical philosophies. One of the first questions they asked me was: "At the start of the primary cycle, are children not too young to tackle the sciences? Are they not still caught in the realm of the magic?"

And indeed, they are. Children's ability to marvel, to sense that we are surrounded by extraordinary things, is a precious gift—as long as it does not frighten them. The complexity of the modern world can be a bit terrifying and can provoke a reaction of discouragement. It is crucial to avoid this pitfall, to make them discover a few basic concepts through first-hand experiments. There exist books of science experiments that are aimed at very young children and are designed to make them experience the joys of discovery.

A simple project, for example, consists in mixing bicarbonate and vinegar in a bottle and placing the bottle flat into a bathtub filled with water. A jet is expelled behind the bottle, propelling it forward. The kids may feel that it is pure magic, but at least they would be creating it themselves, with their own hands. It would not be something they watch passively on television.

One observation which I find particularly distressing in our country (less so in the United States) is the fact that until a rather advanced age, often some two years after graduation, students typically have never experienced the workplace, and have never participated in a task requiring manual labor.

I am absolutely convinced that young children can be introduced to scientific knowledge as early as the primary level, and that it can be done simply and cost-effectively. As far as my discipline is concerned, one could involve the French Physical Society (French acronym SFP), a professional organization made up of people full of good will. Not long ago, this society established a communications office. I would suggest that this office set up an answering service for the purpose of centralizing calls from teachers in search of some documentation, or of a diagram for an experiment, and the like. A staff person with access to a good address book could categorize the requests and forward them to the appropriate researchers capable of answering them. I am more inclined to believe in the educational benefits of such an operation, if well run, than in many didactic reforms destined to end up in the dungeons of the National Education ministry.

A Cultural Deficit

There is a great need in France to open new communication channels for scientific information. I am not certain that the traditional media can provide the ideal vehicle to raise the level of general knowledge. As things stand, the scientific proficiency of the generalist press and of the audiovisual media in France seem to me rather weaker than in Japan and in Anglo-Saxon countries. To deal with scientific information is a craft which requires talent and a particular type of training. Astrophysics has managed to find effective spokesmen. I am somewhat familiar with the case of Carl Sagan, professor at Cornell University, in the United States, who has made the popularization of astrophysics his second profession—with great success. In France, Hubert Reeves has taken up the challenge with more modest means, but with much talent and by investing a great deal of his time. Those are positive cases, but all too isolated.

As for me, I decided three years ago to step in the ring, in other words to show my face in front of cameras and to speak into microphones. But I had a slightly different purpose, related primarily to the somewhat precarious state of the Institute of Physics and Chemistry, for which I am responsible. I do not regret my efforts: the media allowed us to make the public more aware of the school and of its unique style.

From a general standpoint, I would argue that the so-called educational television programs, including the best ones, can dangerously distort the viewer's frame of mind. My concern applies to all kinds of subjects, but the example that comes to mind is a television series on Romanesque architecture. The cameraman was showing the basilica of La Madeleine in Vézeley, particularly zoomed-in shots of details of the cornices. The details were magnificent, of course, but the viewer no longer had to go discover them for himself. To be fed images, even carefully selected ones, encourages passivity. The same is true in science, and that is why I believe that children should be encouraged to try even simple experiments, like the rocket bottle in the bathtub I mentioned before.

Ecology and Ignorance

Another concern of mine, of which I was not even aware before I started my round of lectures in high schools, is the realization that our citizens do not possess the basic scientific knowledge to enable them to make sound economic and ecological decisions. A democracy cannot function without educating its citizens. It is our duty to provide our younger generations with a sufficient store of scientific information.

Two of the most pressing problems faced by our world are the birth rate and the environment. But it appears that the information dealing with these issues is both mediocre and dangerous. Everything is approached from an emotional angle. Between an almost religious belief in the virtues of the natural and the fear inspired by some images intended to shock, there is no room left for an objective judgment. Most people have not learned to deal with orders of magnitude, to evaluate the relative and absolute importance of a phenomenon, and to estimate the cost of an alternative solution. I am not proposing to impose on our students a detailed and thorough knowledge, but to instill in them an appreciation of the problems based on common sense and on a few elementary facts concerning matter, its organization, and its behavior.

Environmental problems are often managed by specialists in "simulations," that is to say, people whose competence is more in the area of computer programming than in interpreting scientific data. Large computers can produce predictions that look quite credible even though the numerical inputs may be deficient. That is one of the great evils of our time. Unfortunately, many people firmly believe that the computer tells only the truth and predicts the inevitable (the same kind of belief prevailed in the nineteenth century toward printed text). The expert engaged in computer simulations must be right since his hardware is endowed with a power and speed of calculation far surpassing that of the human brain. The strength of numbers bolstered by the power of images is enough to sustain in the public an irrational, quasi-mystical mind set.

I now come to the much-talked-about greenhouse effect. The scenario goes like this: the concentration of carbon dioxide (CO_2) in the Earth's atmosphere has been increasing for at least the past century, possibly as the result of human activities. This additional CO_2 may absorb solar radiation in the atmosphere, warming it, and in turn warming the surface of the globe changing the climate everywhere; in addition polar ice melts, swelling the volume of water in the oceans and flooding low-lying continental zones. But while the increase in CO_2 is clear, the warming trend in the Earth's atmosphere—the "greenhouse effect"—is less so. In fact, predictions are difficult at best. First, the primary component of the greenhouse effect is water. The contribution of CO_2 is a corrective effect coupled to that of water in a subtle way. Second, the behavior of CO_2 in the presence of water is poorly understood. We know that water contains vast amounts of the gas in solution, but we know very little about the nature of the equilibrium between the gas and the ocean, or about the time scale of the absorption-release cycle of the gas (is it 20 years? 50 years?). The truth is that the models used in 1994 to predict the future climate fail to correctly account for even the present conditions! The modelers correct for this deficiency by adjusting the rate of exchanges between the atmosphere and the ocean. These manipulations give the appearance of credible results. But they can also jeopardize the entire prediction exercise, as shown in a recent study done by the Massachusetts Institute of Technology. As a commentator in the journal *Science* asserted: "In climate prediction models, everybody cheats (a little)."

The greenhouse-effect problem must be pursued actively, but many additional studies are needed before reaching a conclusion. Nevertheless, the press often quotes dramatic predictions concerning the phenomenon, based on questionable simulations. Ordinary citizens and their representatives do not have, for lack of a solid scientific education, the means to recognize by themselves these gross discrepancies between a natural phenomenon and a hastily assembled model.

This is not to denigrate the field of ecology. A thoughtful ecological watch must be maintained by researchers and engineers, as well as by citizens. But it is a disservice to sound the alarm whenever a self-appointed expert calls a press conference to announce the imminent end of the world.

More Conscience Calls for More Science

An important question, about which I confess to harboring a great deal of pessimism, concerns the energy dilemma. Should one choose fossil fuels (coal, oil) or nuclear energy? In other words, do we want a civilization that produces CO_2, or one founded on nuclear materials? I am fearful, in view of the current trend in American society, that decisions are being made based on faulty criteria.

Americans currently show a strong aversion toward nuclear energy. The Three Mile Island accident has left its imprint on public opinion.

A slight revival has, however, taken place of late, perhaps in response to a number of mishaps involving oil tankers, but also because the American Academy of Sciences published a report suggesting that the risks of nuclear energy are less serious, assuming that this option is managed correctly, than those associated with a massive release of CO_2 in the atmosphere. Unfortunately, the operator of a Russian power plant leaned a bit too heavily on vodka and blew up a reactor, and that was enough for the American public opinion to make an about-face. The choice between thermal and nuclear seems to depend on the latest news flash.

The public overestimates the dangers of nuclear technology, at least with respect to power plants built in Western nations. We tend to forget the time when coal extraction killed so many mine workers. Firedamp explosions still occur, gas fields still blow up, and miners are still stricken by silicosis, but these tragedies now affect primarily the third world, and they receive hardly any attention. As for oil, it generates its share of pollution, tensions, and armed conflicts. Nuclear power, on the other hand, suffers from its association with the bomb and the destruction of Hiroshima and Nagasaki and from the more recent calamity at Chernobyl.

I am defending neither the oil industry nor the nuclear lobby. I simply wish that the public would gain a scientific maturity sufficient to grasp the stakes of the debate. The nuclear option is far from being immune to dangers, the most serious of which are not necessarily the ones which we are kept informed about. In France, the Atomic Energy Commission and the national electric utility have demonstrated that it is possible to build an inventory of reliable nuclear power plants capable of producing electricity at a competitive cost. But in trying to extrapolate this data to the third world, one cannot ignore the risks of nuclear arms proliferation. Modern reactors can easily be modified to produce plutonium. The temptation is great to adapt this civilian technology to military purposes.

To prevent this possibility, should one reserve the nuclear option to developed nations and deprive the third world under the pretext of its political instability? I personally do not believe that the third world, given its birthrate, can forego nuclear energy. This issue is closely intertwined with the fundamental problem of population growth.

Researchers are trying to come up with a reactor design in which the fuel would be in the form of encased micropellets, impossible to reprocess to extract plutonium. But the execution of a nuclear program of this magnitude will require one or two decades. In the meantime, Pakistan and North Korea, to mention but two, will have manufactured more weapons than are necessary to declare war on their neighbors.

What about alternative sources of energy? With sustained efforts, we are sure to succeed in converting solar energy into electricity with a reasonably high efficiency. This solar card is viable in certain warm, sunny countries, but it is unlikely to be the ultimate solution. It will, at best, be a supplementary source of energy, if properly targeted. Israel would be a perfect candidate. The Jewish state offers tax incentives for simple solar systems producing hot water for residential use. But even in this case, despite favorable climatic conditions and an affordable technology, solar energy remains only marginally competitive. The problem of energy storage has not been satisfactorily resolved. The sun provides energy during the day, while the demand for energy spreads over 24 hours. The cost and sheer bulk of storage batteries, and the chemical pollution they themselves create, combine to limit the widespread use of solar energy.

Will we some day succeed in installing huge solar absorbers in the Sahara desert? I am rather skeptical about the prospect. Several unknowns remain, including a climatic risk pointed out to me by Pierre Aigrain. If a totally black surface area of 1 square kilometer is set up in the desert, this zone will be hotter than the surrounding environment. It will act much like a chimney, creating gigantic air drafts, likely to cause turbulences resulting in wind storms. Nothing is simple!

Toward tidal energy, which relies on the amplitude and the driving force of ocean tides, my feeling is one of sadness. A tidal power plant has been built in the bay of La Rance, near Saint-Malo. More ambitious projects, like the bay of Mont-Saint-Michel, and the Bay of Fundy in New Brunswick, Canada, have been abandoned. These plants are an absolute nightmare for engineers. Operating turbines in a stable manner with a fluid as corrosive as seawater, while coping with the vibrations caused by waves, involves a cost so exorbitant as to preclude commercial exploitation. Production remains low, only 540 GWh (gigawatt-hour) in France. It is a great disappointment, because the principle was so attractive.

Another avenue has generated great hopes for the distant future: that of nuclear *fusion* [9]. For a long time, I believed it necessary to invest in this possibility. Lately, I have lost some of my faith. Fusion poses formidable technical problems. How does one manage to confine what is, after all, essentially a hydrogen bomb within the walls of a hangar, no matter how large? Nobody has the answer. Here again, the information released to the public at large is tightly controlled by special interest groups.

What is certain is that ecological problems will not be solved by a return to the past. A construction worker will never relinquish his jackhammer in favor of a pickax, nor will the housewife give up detergents to resume making her own soap with ashes! Stopping the production of aluminum would not serve the interests of the residents of the valleys of the Alps, although they do complain, and rightly so, about the smoke generated by that industry. The solution is to invent a new process to eliminate the offending smoke. When it comes to getting rid of pollutants, the future belongs to chemistry, not in the bankruptcy of an entire industrial sector. This is, once again, a matter of common sense.

The public sometimes allows itself to be seduced by chimeras. There exists in the United States a movement demanding a ban on plastic packaging, which is perceived as more polluting than paper. Opponents of plastic are presumably driven by a noble motive, but they have no sense whatsoever of orders of magnitude. The manufacture of plastic packaging requires an oil refinery and an adjoining plastic factory, which amounts to an industrial site covering an area of a few square kilometers. Three such sites, representing a total area of 15 to 20 square kilometers devoted to the production of plastic packaging, can satisfy the entire demand of the United States. If a decision were made in favor of an all-paper substitution, the same demand would require planting trees over a territory covering *five North American states*—five states emptied of their population and devoted to the exclusive exploitation of lumber. Not to mention that the paper industry itself has historically been one of the greatest polluters of our planet (cellulose fibers are treated with very aggressive chemicals). Scandinavian countries have experienced the problem firsthand: the Baltic sea has been badly contaminated by these industrial effluents. At the present time, the manufacture of paper has been greatly cleaned up, thanks to a vigorous effort of industrial research. But the public totally ignores the very real problems of paper.

Plastics, on the other hand, constitute pollutants too, particularly in terms of their disposal. Here again, research should be in a position to provide incremental solutions:

- Recycling is one solution, naively supported by public opinion. It is usually wrong and far too costly. Properly incinerating polyethylene, for example, and recovering a small amount of energy is preferable to recycling. In certain specific cases, a well-thought-out recycling operation can be considered. For example, polystyrene is soluble in crude oil. Perhaps discarded polystyrene could be inserted into the normal operation of a refinery, which could re-transform it into basic organic products.
- Biodegradable plastics will be used primarily in the countryside, where large surface areas are available to let them be digested by soil bacteria. This does not solve the problem of urban wastes.
- Photodegradable plastics, which are decomposed by the sun's light, could become instrumental in curbing pollution in streets, on beaches, etc. At present, they are expensive. Obviously, this is a research area to pursue.

PART III

a duck feather
... scientist is neither a prophet or an Einstein-like savant ...
the cunning boat

CHAPTER 1

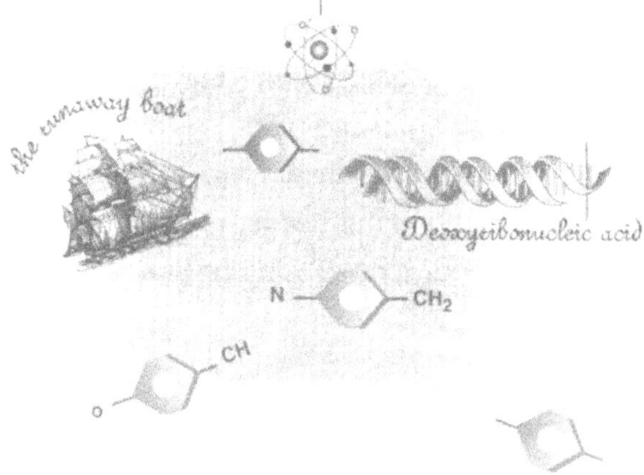

A Pedagogy for the Real World

Good Teachers

I was a high school senior in 1948. This was practically a Paleolithic era, when high school students were very few—the number has since increased tenfold. I was fortunate to have absolutely outstanding teachers in all subjects, people who today would be university professors. Our training was first-rate.

After completing high school, I entered an undergraduate preparatory program which, for reasons unclear to me, no longer exists: Normale Sciences Expérimentales.[1] Fate put me on board the right train. This program was quite different from the traditional preparatory classes emphasizing math as a major: it included mathematics, physics, chemistry, and biology. This last discipline offered the most opportunities for discoveries, and demanded the greatest effort.

[1] Translator's note: This program led to the experimental sciences branch of the École Normale Supérieure.

We emerged from these two years with a vision of the world that was far less abstract than that of the preparatory mathematics majors. In physics, we were taught to adjust an interferometer, and we learned how to perform quantitative analyses in chemistry: you were given a soup containing unknown ions and your assignment was to identify them through a systematic procedure. From the point of view of experimental skills, observation, and deductive abilities, nothing can be more formative than this kind of training (it also happens to be inexpensive to implement).

I was also privileged to have had the chance to spend some time as a summer intern in a marine biology laboratory (the Arago lab, in Banyuls).[2] Every student should dabble in biology, even if he has no intention of ultimately becoming a biologist. To begin with, this science has the virtue of offering a panoramic view of the living world and of its stunning diversity; a firsthand observation of nature's richness of imagination. On a typical day, we would go hunt for a small scorpion in the mountains, catch it without damaging it (which requires a certain dexterity), and immerse it in paraffin before placing it in a stereo microscope (a magnifying glass would be adequate) to dissect it, using two pins as scalpels. Next, we had to describe everything we had seen with a series of drawings. This type of hands-on activity involves important and varied skills, like visual acumen, capacity for observation, coordination, and appreciation for careful work.

This object lesson came, unfortunately, a bit late in my life, when I was already 18. I wished that somebody had taught me to make decent drawings three years earlier.

Once accepted at the École Normale, like any other student who has just successfully passed an entrance examination, I thought that I had it made. So, I spent the next six months doing nothing, or, more precisely, getting busy with projects that had very little to do with academics, like building an audio amplifier and organizing the school dance. If prolonged to excess, this decompression phase can be costly! The

[2] Banyuls is located in the foothills of the Pyrenees mountains, on the Mediterranean coast.

ability of neurons to store new information evolves with time. To fail in a subject and hope to come back to it later is a losing proposition. Later on, the mind is not as sharp, and other priorities interfere with the resolve to pick up subjects that were set aside.

Fortunately, the École Normale had on its faculty extraordinary educators. Among them, three strong personalities had a profound influence on me.

- Yves Rocard, whom I have already mentioned, who had participated in the development of the vacuum tube, in the field of mechanics in the 1930s, and even in bridge construction: an eclectic mind, pragmatic and humorous.
- Alfred Kastler, brilliant opticist, whose delivery was so illuminating and clear.
- Finally, Pierre Aigrain, the youngest of the three, founder of French semiconductor research, with a bubbling imagination both on the experimental and the theoretical levels—a genuine hurricane. His students wonder to this day how they survived intact, but they keep an unforgettable impression of him.

I also had the pleasure of spending one summer at the École des Houches.[3] Physicists owe a great debt of gratitude to Cécile Dewitt, the courageous young woman who had succeeded, shortly before I got there, in creating this school and turning it into a unique cultural beacon.

Later, as a teacher, I was assigned relatively easy courses. I did not have to teach students in their first two years of university, which are typically the most exacting courses to teach. Instead, I taught classes at the graduate level, to a select group of students; later, I taught at the Collège de France, which caters to a varied audience, but one that is always highly motivated.

The facts that I belong to the postwar "spoiled generation" and that I taught in institutions spared from turmoil do not confer on me much

[3] Translator's note: A famous yearly summer school held in Les Houches in the French Alps, during which a small, international group of experts and students gathers for a few weeks of intensive lectures on the latest developments in advanced theoretical physics.

authority to talk about the problems of education. I was fortunate to have had good teachers, and, later, good students. Nevertheless, I feel the need to address in these last chapters a number of concerns and proposals concerning education.

Too Much Ignorance

I am not happy with our educational system, its specialization options and its selection criteria. This negative judgment is targeted only at the areas close to me, namely, science and technology. But it does involve issues which, of necessity, spill over from the sphere of secondary schools. My view is not an indictment of the competence of teachers (who are, for the most part, aware of the situation), or of the good will of the students. It relates to a system, its mentality and its stifling patterns. I have few illusions on the possibility of remedying the problems with broad reforms. In the light of earlier chapters, the reader will have understood by now that I am more inclined to bet my money on pilot programs run in the field, with specific objectives and modest means. But certain things need to be said, as unfair or harsh as they may sound to some. Science is too serious a matter—particularly in terms of how our youth approaches it, its professions, and its social implications—to be left at the mercy of the system's inertia.

I believe that there is too much ignorance all around. A common attitude is to say: "Science is dangerous; it is an endeavor which serves only industrial interests, and our children should be protected from it." This is not a realistic posture. Our children will require enough of a scientific and technical background to survive even if they don't become scientists by profession. To shield or discourage children and adolescents from a scientific culture is simply criminal, because it leaves them unable to cope with the problems that await them. They must become acquainted with a concrete science, very different from that to which they are exposed now.

Over 20 years ago, one my teachers, Alfred Kastler, wrote: "In our educational system, the initiation to the sciences is completely dominated by a mathematical focus increasingly oriented toward the abstract.

The objective is to develop in children exclusively the sense of logical deduction. Manual skills, visual acumen, the sense of observation, an interest for the physical world which surrounds us, are all qualities that are neglected and downgraded. A student's aptitude for the sciences is judged on the basis of his taste for mathematics and logic." The observation retains all its validity to this day.

Manual Work

In front of an audience of French high school students, I emphasize ordinary, day-to-day facts, while for young Americans my speech would focus much more on theoretical concepts. The two systems are diametrically opposed. Americans undergo relatively little mathematical training in high school. Even the style of lectures is different. French students hesitate to assert themselves, to ask questions. There is no such inhibition in the United States. The French student may be more knowledgeable than his American counterpart (who would probably fail to measure up to our scholastic achievement standards), but the young American is more self-confident, more assertive, more inquisitive about the real world. These are two educational styles, each with its shortcomings—programs too heavily focused on theory on our side, not enough on theirs. But Americans have a higher education system of outstanding quality, which allows students to bring themselves up to international standards, while the French model only aggravates the problem. I am here thinking specifically of this uniquely French tradition: the entrance examination to institutes of higher learning, which has no equivalent in Germany or in Anglo-Saxon countries. I will return to the subject of entrance examinations later.

In comparing French and Japanese students, one distinguishing feature clearly sets them apart, and it is not to our advantage. I am talking about learning how to write. In Japan, a child takes seven years before he knows how to correctly draw characters. Far from being a handicap, this painstaking learning process develops in him an essential quality: the taste for a job well done. In our primary schools, this aptitude is increasingly neglected, if not downright discouraged. Take the example of a youngster in kindergarten who works hard to create a pretty drawing, while his classmate sloppily rushes through his. So as not to traumatize anybody, the teacher will declare both pictures equally good and will refrain from encouraging the scribbler to apply himself more. The latter will then quickly learn that one can get away in life with sloppy work. An important difference between Japan and France has to do with this respect for quality. Our cleverness can never make up for this manual skill, this ability to attend to details. The Germans share with the Japanese this sense of a job well done. I am not sure where this trait, which has made the reputation of their mechanical industry (among others), comes from. But I only need to look at a masterpiece by Albrecht Dürer (1471–1528) to recognize in this German Renaissance painter and engraver an extraordinary level of craftsmanship, in its detail and realism, culminating in a work of art of exquisite perfection. An Italian equivalent might be Giovanni Bellini (1430–1516), who, incidentally, knew of Dürer and of his work.

The French overestimate the power of the mind. To be sharp and brilliant in deriving a proof cannot be the only criterion for competence. We also need craftsmen, artists, people of sensitivity, as well as perfectionists. We tend to brag about our cleverness and resourcefulness: how smart is one who can dispatch a mathematical exercise in a trifle. We nurture the cult of this form of intelligence. Without a doubt, it is useful to quickly dispose of small mathematical problems; it does save time, but it is somewhat superficial.

In order to rediscover the sense of quality, work ought to be, in my opinion, the primary goal of education. It is said that we are moving toward a leisure civilization, and it may be true, although our inevitable amalgamation with the third world does not allow us too much loafing. Yet, if we do spend less time on work, it is all the more reason to do it well.

Since the 1968 events, students have become an influential group. A great deal of power is in the hands of young voters and consumers who, for the most part, have never come in contact with the workplace. A few do work in the summer, but as a general rule our young people take long vacations, with little formative value. Yet it ought to be possible to encourage small enterprises to accept students as summer employees. To work in a garage seems to me the best initiation to a professional life. One gets to learn techniques that are anything but trivial, and one has to build relations with co-workers, with the boss, and with customers. This kind of training helps to inculcate a sense of a job well done, but it upsets entrenched habits. One would have to change some aspects of regulatory legislation affecting the workplace, and also convince garage owners to offer their summer trainees more than just a broom to clean up the floor (the risk is very real!). If necessary, put in place a control and follow-up mechanism, and avoid penalizing the apprenticeship programs that train the future professional mechanics.

On the other hand, I am rather skeptical about the merits of school groups visiting large factories or ultrasophisticated research centers. A genuine dialogue can only come from working side by side. For the same reason, I doubt the usefulness of group travel. To send off a pack of youngsters on the roads of Greece or England, 40 to a bus, seems to me to be completely devoid of any pedagogical virtue, offering no personal contact with the country, its mentality, and the knowhow of its people. If we want to send our children abroad, it should be for the purpose of working there. It is only through shared work, not through casual tourism, that one comes to know a country.

The Sailor's Bar

In rural France of yesteryear, children used to be in daily contact with nature and with the world of craftsmen, which gave them a sense of observation and of manual work. These reference points have all but vanished in our urban civilization, and we must find substitutes.

Access to computer technology is a necessity, but if we are content to sit our young people in front of computer monitors (which they love), we are at risk of losing something precious. To form a generation that knows only how to hit a keyboard and produce reports is, to me, a scary prospect. The time spent on this type of development should be sharply restricted and balanced with a more practical culture, preferably one based on activities conducted in the outdoors. I am a great advocate of botanical or geological walks. To study the flora, to identify the organization of a landscape, to look for fossils, to examine minerals, and so forth teach one to observe the real world, to discover objects, to interpret phenomena. This sort of activity is not very complicated to organize: all it takes is a bus and a set of very simple tools; the ambience is typically quite congenial; the day often ends around a café table, and everybody benefits.

To come back to young American high school students, one of their strengths is to have learned to roll up their sleeves at a fairly early age. From the time they are about 12 years old, they mow lawns, wash cars, etc. It is a firmly established tradition in the United States. Later on, many finance part of the cost of their education by working, sometimes hard. I have a friend, now a successful scientific leader, who paid for his university education by spreading concrete at construction sites in Boston. I also know a young woman, an assistant district attorney in Rhode Island, who earned a living as a bar waitress during her studies. This kind of experience is not irrelevant when dealing with criminal matters. How many French judges have immersed themselves during their student years in an environment other than the Law School they attended? The administration of the École Nationale Supérieure de la Magistrature (ENSM), concerned by this deficiency which I discussed during conferences, assures me that their students routinely go through summer internships, notably in hospital administration.[4] Although this effort is commendable, it is one thing to be an administrative trainee, and quite another to work in a sailor's bar. Too many parents have a tendency to be overprotective of their children because of the looming prospect of the entrance examination that awaits them. Summer work, or a modest student job that provides some degree of independence, a real-life experience, and an opportunity to exercise skills that have nothing to do with academic aptitude, has no room in the context of such a loaded hurdle—particularly if the job involves manual labor. I, myself, regret that I took the easy road in my youth, when I earned some money as a tutor. Mine is an example not to follow!

[4] Translator's note: ENSM is an Institute of higher education, specializing in the fields of jurisprudence and law.

EXPERIMENTAL COMMON SENSE

A century ago, the sons of farmers might not all have known how to shoe a horse, but they knew how to sharpen a scythe and how to use it. They had a practical knowledge and a sense of observation which are sorely lacking today. I do not at all advocate a return to nature, but I wish that the virtuosity displayed by engineers and scientists in handling equations would match their ability to deal with concrete problems. Since theoretical modeling with paper and pencil has occupied a good portion of my life, I am in no position to disavow it, but I see a problem when it becomes the cornerstone of a scientific education. Ignorance of the real world causes grave distortions.

When young people ask me to guide them in their orientation, my answer is that too many specialize in mathematics and management, which remain, after all, rather cold disciplines on a human level. Many so-called scientific schools mold engineers whose primary ability is to digest reports as quickly as possible and generate new ones which, in turn, will be analyzed in record time. To become a machine absorbing and exploiting written reports is not, from my perspective, the ideal path toward a balanced life. As a counterexample, medicine, with its very wide spectrum of outlets, from biological research to clinical practice (which truly qualifies as an artist's craft), seems to me to be a more attractive avenue.

Behind these excesses often lurks the worship of the tool. Each era has its favorite gadgets. In Laplace's time, it was differential geometry and differential equations. Nowadays, it is the computer model, with all the trappings of digital technology. An individual steps forward with a mathematical model and claims to have unlocked the underlying rules of an economic problem, without ever having acquired a solid practical knowledge of exchange networks, of the dynamics of an enterprise, and of its administration. I find this quite frightening. In more physical subjects, I often come across jackasses of that sort. So and so swears to be able to explain a particular phenomenon, only to see his theory cave in after 15 minutes of discussion, simply because he has not bothered to invest the few years necessary to truly understand what he is talking about.

To be sure, computer technology is indispensable. Many experiments are automated in our labs, which obviates the need to spend nights monitoring their progress. Still, simulation at the molecular level, to take one example, depends critically on the common sense of the individual who is doing the work. It requires teams with a broad range of talents and with intellectual honesty.

Equations are unavoidable in physics. One also cannot do without detailed numerical studies and simulations of phenomena that would be too long or too costly to study directly. In short, no anti-equation crusade is called for, but, at the same time, no pro-equation dictatorship either.

When I welcome a freshman class at our Institute of Physics and Chemistry, I insist on concepts that most math majors are not at all familiar with, and which, in my view, are given too little emphasis in our educational system. This "reeducation" phase takes a minimum of two years. I am constantly astounded by how little common sense recently degreed engineers have. I often tell them about a test that Enrico Fermi used to give his students at Columbia University, in New York. (Fermi, who has already been mentioned in this book, was a great theoretician in quantum statistics, and nuclear physics, as well as a builder of machines, such as the first atomic reactor and the first particle accelerators.) On the first day of classes, he would ask them: "I want you to estimate the number of piano tuners in New York City." The reasoning that he wanted them to use was extremely simple: New York has 10 million inhabitants, with roughly one piano for every 30 families. With an average of three people per family, that's 100,000 pianos. How long does a piano remain in tune in a climate as rough as New York's? Let's say three years, or 1000 days, which means that 100 pianos need to be tuned each day. Assuming that one piano tuner can manage two instruments a day, fifty of them could handle the job. With a margin of safety to account for weekends and travel difficulties, one gets in the neighborhood of one hundred. Next, Fermi would ask his students to consult the yellow pages of the telephone book to verify the actual number of piano tuners. The order of magnitude turned out to be quite correct! That is the kind of commonsense education which, in my view, we must encourage.

Ekman and the Drift of Ice Floes

Nothing has more descriptive power than examples taken from history. Among the most inspired experimentalists, I often quote the case of a Scandinavian oceanographer by the name of V.W. Ekman. At the beginning of the century, the young Ekman was watching ice floes drift in the Baltic sea during spring thaw, pushed by a strong westerly wind. The ice pack moved in the general direction of the wind, but after careful observations, he noticed a remarkable detail: the drift did not coincide exactly with the direction of the wind; it was slightly to the left. Back at his house, he thought about this and finally convinced himself that the small discrepancy was a manifestation of earth's rotation, transmitted through a fluid medium. He proceeded to work out what is now called Ekman's theory of layers, a prerequisite to our understanding of the great oceanographic currents. What is noteworthy in this discovery is that calculus played quite a trivial role. Anybody could have handled that part. The stroke of genius was to have noticed that the ice floes were not drifting precisely in the expected direction.

Leonardo da Vinci, Engineer

An example which I discussed recently in a class at the Collège de France involves Leonardo da Vinci (1452–1519). A mind of universal scope, Leonardo became interested in the phenomenon of friction: what is the force required to move an object resting on a support? He conducted experiments with inclined planes, where the force acting on the object increases in relation to the slope of the plane (Figure 52). If one decreases the slope, one observes that, below a certain threshold, the object no longer moves. The corresponding applied force is called the threshold force. This led Leonardo to the discovery of a stunning law, almost incredible in a sense. He placed on the same surface the same object, but on different sides, which means different contact surface areas and different pressures. To his amazement, he kept finding the same threshold force! He had uncovered an extraordinary law of friction, which was explained only recently by F. Bowden and D. Tabor of Cambridge University.

Figure 52

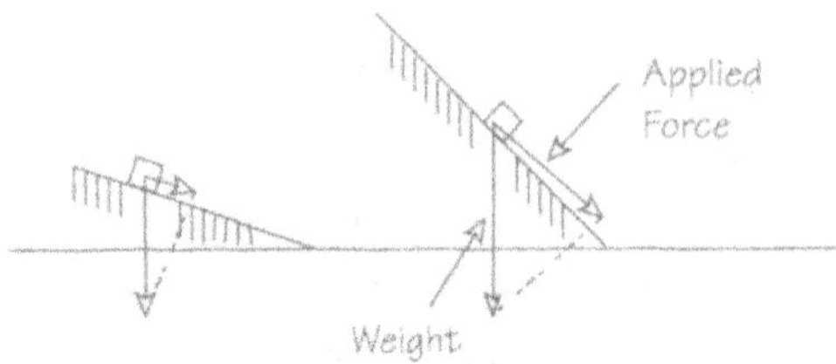

Galileo's experiments on friction.

The discovery of this law (among many other things) by Leonardo da Vinci is remarkable. Sadly, it was completely forgotten for a century and a half, before being rediscovered in 1699 by Guillaume Amontons (1663–1705), and published by the Academy of Sciences. The Academy did not, however, publish before taking some precautions: the academicians deemed this law so bizarre that they asked another engineer, named La Hire, to redo the experiments to convince themselves of its validity.

That is the experimental style which must be reemphasized, quite along the line of Benjamin Franklin: economy of means, inventiveness, and precision.

In our advanced societies, the temptation is great to crush hazelnuts with a hydraulic press! Unlike our predecessors, we sometimes rely too much on heavy hardware, or on huge programs.[5] We must give our students reference points to steer them away from the tendency to systematically call enormous equipment to the rescue. This goes for the applied engineer as much as for the fundamental researcher.

The "Auguste Comte" Prejudice

I now come to a prejudice typical of French culture, inherited from the positivism of Auguste Comte. This nineteenth-century philosopher achieved some degree of fame by inventing a classification system of the sciences. At the top of his hierarchy was mathematics; at the bot-

[5] Freeman Dyson, one of this century's great theoreticians, has written an incisive and humorous analysis of several large programs (space, high energy, etc.) in an excellent book: *From Eros to Gaia.*

tom was chemistry, which according to him "barely deserved the name of science" (!); in the middle were astronomy and physics. This classification dismissed out of hand geography and mineralogy, sciences which were declared concrete and descriptive, retaining only those that were theoretical, abstract, and general. The tone was set! It is ironic that this philosophical concept came from an individual who had once written in a letter "The only absolute truth is that everything is relative," and who claimed to be steeped quasi-religiously in factually observable laws, in other words, laws verifiable by experiments. The "Auguste Comte" prejudice corrupts to this day the teaching of the sciences, the scientific disciplines, and even the scientists themselves. It also contains the seed of contempt for manual labor, which has interfered for years by curbing every attempt at reform to revalue the manual trades and their apprenticeship. These reforms have invariably been blocked by numerous conservative groups and parents hostile to them on principle. As a result, the attitude toward manual work has scarcely evolved at all.

We used to have in Paris an excellent mechanical school emphasizing practical training: the School of Arts and Trades. But I am fearful that even it is drifting toward a more theoretical style.

An example comes to mind, of some graduates of the Polytechnic School of Paris attending an advanced program at Orsay to learn solid-state physics.[6] They would often show up convinced that they knew everything on the basis of calculations. On their final exam, I would give them a problem of the following type: "Imagine a thin, evaporated metal film, like lead, 1 micron in thickness. A cosmic ray with an energy of 10 MeV (megaelectronvolt) traverses the film, which is held at a temperature of 4 kelvins. A voltmeter is connected between the edges of the film. What are the amplitude and duration of the resistance pulse that can be measured across that film? Is it possible to use this design to build a simple cosmic ray detector?"

The students would go off and think about the problem for an hour in front of a blackboard. The solution was rather elementary at this level

---

[6] Translator's note: The Polytechnic School of Paris is a famed military institute of higher learning, founded under Napoleon, specializing in abstract mathematics.

of studies. One starts by considering collisions between charged particles to evaluate how much energy a fast proton of the cosmic ray gives up to the electrons of the lead film; this determines the energy input. To specify how the energy involved diffuses, I would add: "This lead film contains 1 percent impurities," which, the student should know, implies that an electron travels about 100 times the distance between lead atoms before it gives up the energy it acquired when it was hit by the cosmic ray. Two or three general concepts of this kind are all that are needed to predict the amplitude and the duration of the thermal pulse. But the typical Polytechnic graduate I inherited at the time would remain stumped in front of his bare blackboard. One of them finally blurted out (I will never forget his comment!): "But, sir, what Hamiltonian should I diagonalize?" He was trying to hang on to theoretical ideas which had no connection whatsoever with this practical problem. This kind of answer explains, in large part, the weakness of French industrial research.

Among all the catastrophes brought about by the positivist prejudice, none is worse than the widespread contempt for chemistry. I have already pointed out the importance of this discipline for our industrial future, the importance of chemists, these marvelously inventive sculptors of molecules, to whom the French teaching establishment does not do nearly enough justice. An undergraduate math major once told me about a teacher who, on opening day, announced: "I personally dislike chemistry, but I have to talk about it. So, I will start by giving you two hours of chemical nomenclature: what the name of an obscure and complex molecule is, and the like." At the conclusion of the two hours, the entire class was turned off chemistry for life!

When Lucien Monnerie, the director of studies, and I took over responsibility for courses at the Institute of Physics and Chemistry, we had to wage a determined battle to overcome the antichemistry prejudice. Just before our arrival, the students had organized a strike: they all wanted to become physicists. Slowly, we climbed back up the slope with a series of measures: changing labels, opening up several new channels, turning the entire curriculum upside down, and launching a verbal propaganda campaign. It was rather easy for me to sound persuasive; being a theoretical physicist, nobody could accuse me of protecting my own turf. But it took us 10 years to restore the proper balance.

To anyone who wants to form a more precise idea of chemistry, of the life of a typical chemical engineer, I would advise reading the magnificent collection of essays by Primo Levi, *The Periodic Table*. They recount real-life stories. They possess an authenticity and a vitality which give a universal impact to the account of an ordinary fact, the description of minute events. It is an excellent antidote to the poison spread by Auguste Comte's classification scheme.

CHAPTER 2

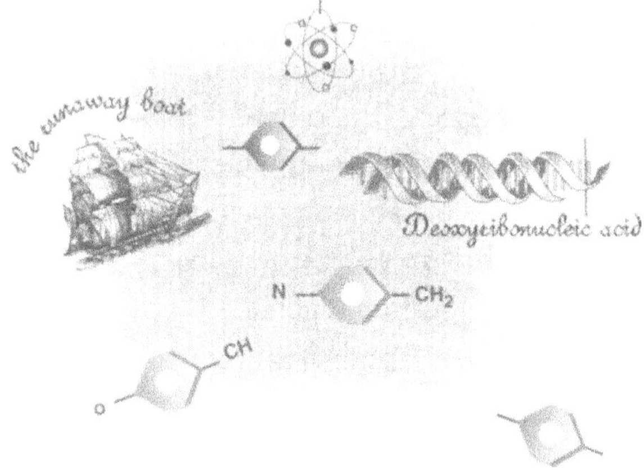

The Imperialism of Mathematics

THE ENTRANCE EXAMINATION THEOREM

There exists a theorem that states: "Whenever an entrance examination is instituted in a scientific discipline, it invariably becomes an exercise in mathematics." I have witnessed this theorem in action when a selection process by entrance examination was put in place in medical school. Within six months, mathematics had phagocytized the system. Before that time, premed studies encompassed a very flexible program of biology, physics, chemistry and math. Students operated in a kind of academic lottery, in which professors would choose to teach what they considered important, a little bit according to their own taste. Sometimes they made the right choice, sometimes not. The hallmark of the first year of medical school was a certain creative effervescence. But from the moment an effort was made to regulate the number of future medical doctors by means of an entrance examination, everything changed. Students started plodding through the same mathematics exercises ("superelementary mathematics" would describe them best). The physics exam became an exam in applied math. The chemistry exam suffered the same fate, requiring complicated calculations of equilibrium chemical reactions. Even biology degenerated into a miasma of combinatorial analyses involving the genetic code. Within a couple of trimesters, the institution of the entrance examination had managed to turn the entire premed arena into a desert. The system of entrance examinations, with its inevitable consequence of the dominance of math, has contaminated the two years of preparatory program which constitute the only access channel toward institutes of higher education in France.

Why is there such a focus on mathematics? It is certainly not because of pressure from mathematics researchers. I believe that the answer lies in a convergence of interests which makes this system practically immutable. On one side, examiners find it more convenient to grade a math problem than they would a discussion of a subtle physical phenomenon where formulas are not the whole story. On the other, students are quite happy with a system which does not require them to store a lot of information, and which is, therefore, amenable to last-minute cramming. In this respect, the conservatism of the student population is no less at fault than the weight of corporatism and bureaucracy. As for the parents, their only concern is to see their children find their niche, and they get nervous at any prospect of change. In reality, the trend toward mathematization turns our graduates, our future engineers, into hemiplegics.

The student who has gone through a preparatory program dominated by formal disciplines, who has passed an entrance examination to an institute of higher education and graduates from it, is at risk of remaining disconnected from the real world during his entire life. Such students are likely to lean toward careers that maintain this state of mind, for example, administration or data management. They may have learned to master certain tools, to prepare reports, but they will suffer from crippling weaknesses in those areas that I have discussed previously: observation, manual skills, common sense, and sociability.

A System Working in a Vacuum

I do not want to be misunderstood. I have nothing against mathematics as a science. It is a superb discipline, of which I am personally very fond. For example, I have followed (from a distance) certain fascinating areas like random surfaces, or knot theory, which touch simultaneously on pure mathematics, statistical physics, and elementary particle physics. What I object to is using math as a selection tool. I am not complaining about mathematics; I merely deplore the fact that it has become the standard against which the level of scientific aptitude and knowledge is measured. This discipline must, of course, be included in preparatory programs, but not to the point of becoming their primary ingredient.

The mathematics taught in preparatory classes corresponds roughly to the mathematics invented by Newton (for instance, differential and integral calculus), augmented by one or two more modern tools like matrix calculus. One year would be enough to absorb this material, rather than the two which are currently the norm. All things being equal, the British educate their engineers in one less year than we do. If we insist on a two-year curriculum, we should strive to lighten the mathematical focus and introduce a genuine experimental background, notably in biology.

The teachers of the preparatory program are not responsible for the current state of affairs. They are restrained in a straitjacket. Their duty is to prepare their students for the entrance examinations, in which grades will be based on ritualized tests. Therefore, they cannot afford innovations.

In addition, its cultural role has made the mathematical examination—and the culture it represents—into a genuine sociological yardstick. It represents to our citizenry what amounts to the "ultimate cultural horizon"—a monument of sorts, an integral part of our national patrimony. But if this system is so productive, and so useful for the formation of generalists, why have the British, the Germans, and the Americans unanimously declined to adopt it? The number of their patents and the quality of their researchers do not appear to suffer excessively as a result.

There are in France 600,000 to 700,000 children per school grade, of which only 40,000 to 50,000 will eventually be admitted to an institute of higher education through an entrance examination. Yet a substantial part of our education system is geared toward this mode of selection. One of the arguments in favor of entrance examinations is their alleged social equity. They supposedly favor neither wealth nor social status. From a historical perspective, the claim was partially justified in the aftermath of the French Revolution, at the beginning of the nineteenth century. Today, to the extent that equality of treatment does exist between candidates to entrance examinations, recruitment is neither more nor less diversified than in British or American universities, in which the selection of students is based much more on application forms. Incidentally, the family dynasties graduating from the Polytechnic Institute, the selection by social background, the high percentage of teachers' children at the École Normale, may evoke some skepticism about the fairness of today's system.

While English youths entering Oxford in the 1930s came predominantly from public schools reserved for the gentry and the wealthy families of the British Empire, the situation has changed considerably since 1945: the United Kingdom's system no longer fits this stereotype. Nor did I observe any socially biased recruitment in the United States. As a matter of fact, the United States is presently tilting toward the reverse excess: at the University of California in Berkeley, the minimum admission grade is about 12.5 for whites and Asians, 11 for Hispanics, and 9 for African Americans.

Lifelong Rights

After a sustained effort of two years or more, French students emerge from their math-intensive preparatory school and enter an institute of higher education. It matters little which specific one they enter, the result is the same: they feel they have reached the mountain top, and they automatically begin a long pause. Why exhaust yourself with work when your future is practically assured? This attitude is doubly troubling. First, I find it outrageous that, just because one has passed an entrance examination at age 20, one should think that one is entitled to life-long social privileges. Other young people who were not as fortunate to earn this passport, either for social reasons or because they did not fit the mathematical mold that I criticized earlier, find themselves relegated to less desirable avenues. Second, it is shocking to me that institutions as richly endowed in terms of structures and support as our institutes of higher education are prone to operating in such a vacuum. During my youth, the Polytechnic Institute was, to be frank, weak in many areas, physics among them. Today, thanks to a smart policy pursued over a period of more than 20 years, it has managed to recruit in most scientific disciplines some of the best faculty in the country. The school now boasts excellent teachers and brilliant students. But there is still too little academic effort on the part of the student body, and few scientific vocations at the end. Graduates are guaranteed a career and quietly wait to land into a suitable spot.

The same problem exists at many other institutes. The fact that the young people admitted to them quit applying themselves just as they begin to be exposed to real problems and to a highly qualified teaching faculty is a shameful waste.

A few schools stray toward the opposite extreme by imposing too intense a pace after the entrance examination. That is the case, in particular, with the École Normale. It once offered a chance to explore according to one's own taste, with few constraints. But in the last 15 years, the load has become much heavier. To graduate with the widest possible spectrum of career opportunities, the science majors in that school accumulate multiple diplomas and undergo an encyclopedic curriculum. In the process, they have lost the time to think. After the furious regimen of prep school, they experience more of the same. Fortunately, their director is aware of the problem; he assures me that things are improving.

On a general level, what I hate most in our institutes of higher education is the proclivity for self-satisfaction—of the teachers, of the students, and particularly of the alumni. It is at the root of conservatism, and of this absurd notion of a "life-long right."

The "Limp" Education in Universities

In France, following high school graduation, the flow of students splits three ways: technical schools, which I will not discuss because I am not very familiar with them; institutes of higher education; and universities.

The university route has problems of its own. The level of proficiency of the average first-year student is mediocre; exams are undemanding, yet the failure rate is high; teachers show little concern for the future career of their students. I am convinced—and I weigh my words carefully—that universities must submit themselves to a serious process of self-examination; they are in dire need of a rigorous self-criticism conducted independently of administratively mandated changes imposed on them from the outside. There are four distinct problem areas.

- The French university system suffers grievously, in my view, from its division into watertight sections. I am appalled, for instance, to see mechanical engineering completely isolated from physics, from materials science, and from biology, and emphasizing, here too, applied mathematics and numerical techniques. This partitioning is harmful. We physicists bear our share of guilt. Each specialty has the weakness to perceive itself as more advanced and smarter than the others (another manifestation of the old Auguste Comte prejudice). Physicists completely ignore chemistry, an intolerable arrogance in our time. Conversely, I am afraid that chemists sometimes paint in their courses a caricature of physics (I think, for instance, of certain courses in chemical thermodynamics). Similar distortions plague physical chemistry. It proved extremely difficult to put in place at the University of Paris a physical chemistry curriculum that was sensible and responsive to the needs of industry: the inertia of corporatism stood in the way.
- Second, the teaching style of universities is flaccid. There is a tacit understanding that the subject of the latest recitation session will constitute the topic of the next exam. Those who superficially bother to keep up with the flow of events are assured of passing. This arrangement constitutes a genuine collusion between teachers and students, each helping the

other cut the branch on which they both sit. The lack of psychological pressure allows the student, with a minimum of exertion, to maintain himself even with the level of exams—but unfortunately not with the level of knowledge required by international competition. Students have the illusion of progressing with little expenditure of effort, but do not realize that they are actually wasting their best years. In this respect, the traditional preparatory schools perform a better job. They do manage to teach how to work under deadlines, how to sweat at a blackboard, and how to handle the pressure of tests. While the university offers a more varied fare and a more relaxed pace, only the most determined, energetic, and self-motivated student can derive any benefit from it.

- Third, there is a lack of connection between the teaching faculty and the industrial pulse of the country. As things stand, we university academics too often provide our students a hazy and obsolete view of research and technology, even though a few points of contact have developed over the years between some of our laboratories and large corporations (which have their own large research laboratories).
- Finally, I deplore the lack of a real recruitment policy for professors on our campuses, in sharp contrast to the extensive, concerted faculty searches conducted by American universities. Our university laboratories tend to select their staff based not on the future needs of the nation but in terms of their self-centered drive to complete their teams. I hope that universities will become more disposed to opening themselves up to the outside world.

Where do theses weaknesses come from?

University units function under a system of collegial administration, which is almost automatically condemned to be weak, based more on maintaining a network of mutual acquaintances than on striving for the collective good of the institution. At my school, at least, as director, I have the power to make hard decisions to improve education, lighten courses, possibly eliminate some or create new ones. Such freedom and authority do not exist within the framework of a university.

All these shortcomings and drawbacks find their way into the final product: the degreed student. Whether he chooses research or industry, he will have a rude awakening when he realizes he is ill-equipped to cope with the pace of work.

For those who want to become high school teachers, the proper channel toward this goal is to go through a university institute devoted to the formation of teachers: teacher's colleges. Will they be in a position to project a lively image of their field? I see two potential problems.

- A substantial part of the student population in these teacher's colleges chooses that route because their career will be assured, not because of motivation. Even if their professors go out of their way to instill in them a tiny flame, I am not sure that it will catch on and resist the assault of time.
- The curriculum in these teacher's colleges is more and more oriented toward the didactic, consisting of rather obscure discourses on the art of communicating knowledge. The would-be teacher is discouraged from taking courses in his own discipline, in favor of those didactic lectures developed internally. There is a long list of fashionable trends with disastrous results in education: in my time, it was Marxism, psychoanalysis, and the "new" math. Today, the latest fad is didacticism.

CHAPTER 3

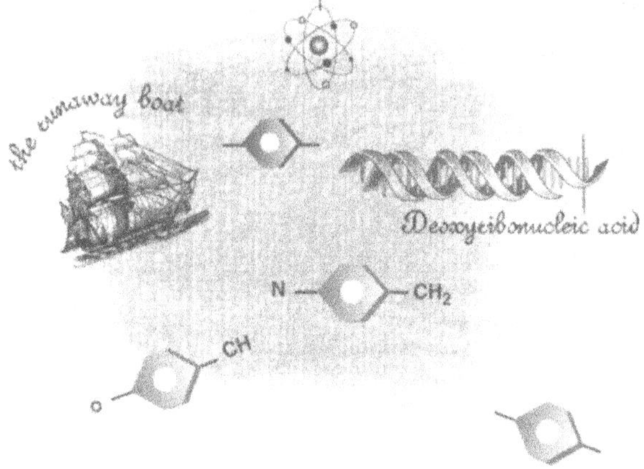

A Little Oxygen

Some Pedagogical Experiments

The options we give our students and the way we teach them must change. What sort of improvements can be proposed? Ideally, I would advocate the outright abolition of the entrance examination system. Because the very concept of the entrance examination is one of the primary causes of the dislocations affecting our education system, my dream is to see it disappear altogether over the very long term. But one must be realistic: it is politically impossible to reform national education by sweeping legislation. In the short term, I therefore favor gradual adjustments. I envision several stages.

As I canvass industrialists, researchers, and academics, most agree that the preparatory programs as we know them disconnect the student from the real world. But the system is so entrenched that no one dares touch it. Therefore, it is best to proceed cautiously with incremental corrections, avoiding structural disruptions. First, favor should be given to courses in physics and chemistry with a more experimental emphasis in preparatory classes. There is a timid movement in that direction, but it is only a limited improvement, one which does nothing to counteract the spirit of the entrance examination, the mathematization of subjects, or the slackening of effort after admission to a school of higher education.

A second stage seems necessary to me. It consists in creating a number of more autonomous pilot preparatory classes, built on the triad of physics, chemistry, and biology. There should be about ten such classes in the Paris region, another ten in the provinces, staffed by dedicated teachers—I am sure they can be found—offering programs developed in consultation with the end users. We nearly succeeded in this effort. But we will have to show more perseverance. Evaluating the impact of such a new program, the response of industry, the productivity of the newly trained engineers, will take about 10 years.

At that point, the time will have come to draw lessons from the experiment, and, possibly, to consider eliminating entrance examinations. It will probably be desirable to retain a written exam as part of the selection process, a form of crude filter to ensure that the candidates possess certain fundamentals. From there on, everything would be handled through oral tests. Here, Claude Allègre, the founder of French geochemistry, has proposed a concept I find intriguing: each student would define one or two subjects for which he has a real passion, unique topics on which he would be quizzed. He would be judged more on his enthusiasm than on the overall spectrum of his knowledge. But the approach has its limits. The danger, pointed out to me by some professors, is that the student, well aware of the system, will, starting in high school, concentrate on only one or two subjects which he will study in depth, hoping that the strategy will make him a sure winner. Any format of competitive selection, even a new one, will prompt the student population to come up with countermeasures.

Another method, which has demonstrated its efficacy in certain great American universities, is the interview. The oral test is replaced by what amounts to a casual chat about everything and about nothing. This recruiting technique seems to me to work quite well, but it is difficult to implement. The evaluators must be motivated, even enthusiastic, and they must be available, because the process is time-consuming.

When it comes to institutes of higher education that rely on traditional math-oriented entrance examinations, I am not optimistic that they have the motivation to spontaneously revamp their selection method. The environment in these schools is rather independent; each school is extremely proud of itself.

My sense is that it will not be possible to bring much fresh air into the preparatory schools, let alone change the system of entrance examinations, without ultimately shaking the coconut tree from top to bottom, from the government to the players themselves. It will take considerable effort and determination.

Giving Universities Some "Muscle"

The university system, for reasons explained in the previous chapter, leaves me very perplexed. Nevertheless, I feel that some suggestions are in order.

- Professors and lecturers should be forced to take sabbaticals, not necessarily to devote themselves to research, but, also, to work in industry. For academics to immerse themselves for a relatively long period, like one year, in an industrial environment would be a significant step forward. There is, however, a problem built right into the system: a young professor is routinely assigned the most thankless courses. As he gains seniority, he earns the right to teach more stimulating classes; he should not lose these privileges by leaving the university for a sabbatical year (his classes could be temporarily covered by substitute professors).
- In order to give students more "muscle," to better position them for successful careers, I would advocate putting more pressure on them through one-on-one tests (with the way recitations are conducted now, an entire class snoozes while a single student is under fire). For example, four students could be quizzed simultaneously in front of four blackboards on four

different questions, under the watchful eye of a professor who could, at any moment, step in to question a response written on the board, and catch the individual on the verge of getting lost. This style of exam makes it possible to establish personalized contacts and to zero in on what the student has not studied enough. Even at the Institute of Physics and Chemistry, we are guilty of no longer using this type of test.

- The ideal solution would be to teach in a style that is more alive and individualized, based on a direct dialogue between the student and the professor. But this approach demands an enormous amount of time, and energy, and a large teaching staff, and therefore money.

With Lucien Monnerie's help, I have established at our Institute a system which we formally labelled *preceptoring* or *tutoring*, after the British model, but in a less perfected form because of financial constraints. At Oxford (and, to a lesser extent, at Cambridge and at Imperial College in London), each student has a scientific mentor whom he consults each week or every other week throughout his studies. The mentor asks a few questions, recommends certain books, and sounds an alarm when the student takes it a little too easy. He plays the role of a goad or a turn signal.

We could not afford a one-to-one ratio and were forced to fall back on a one-to-four format, each mentor taking charge of four students at a time. We hired some mentors as interns at the school and recruited others from among the young researchers based all over the Latin Quarter. This influx was salutary, since our Institute, like many others, has a tendency to withdraw within its walls. The library, well stocked, large, and until then practically deserted, suddenly started filling with people. But, in time, the system proved exhausting because of the one-to-four ratio. Typically, of the four mentored students, one is truly motivated, two are reasonably interested (meaning they are content to recycle photocopies made by their predecessors on the material that they are studying), and the fourth does absolutely nothing. To motivate all four would require biting the financial bullet to achieve a one-to-one ratio.

A pedagogical innovation of this magnitude cannot succeed overnight. The development of a tutoring system took us three years: one year of investigation in England, one year to train the mentors, and one year to implement the system. But it gave us a chance to clean up the teaching staff. Nearly 30 percent of the original faculty was removed. I am convinced that such an overhaul should be carried out every 15 years. It is necessary to prune the obsolete as well as to add new material, of course, because the sheer volume of knowledge increases at a furious pace. In general, it is possible to do a fair amount of trimming in the more mature disciplines because they get streamlined. For instance, the structure of atoms and its relation to the periodic table was often taught in my youth with the help of overly complicated differential equations, lacking any broad overview, whereas the essential concepts can all be grasped with very simple symmetry rules and a few conservation laws.

Too often, education gets lost in the details. In introductory chemistry courses offered at the university, the poor students are forced to swallow a pile of empirical or semiempirical rules dating from the 1920s, which were useful to the spectroscopists of the turn of the century. Nobody needs them in today's chemistry, but they are still taught in universities.

Travel Does Make One Young, but in Later Years

Besides preparatory programs and universities, students can today follow a third avenue, unlocked by the emergence of Europe. This opportunity is not to be overlooked in our disciplines. For instance, there are excellent British universities, not just Oxford and Cambridge, but also Bristol, or Imperial College, London, which provide a much more balanced curriculum than ours do.

While on this subject, I wish to say a word about the union of Europe. Its political ramifications were much debated during the referendum on the ratification of the Maastricht Treaty, as well as during the last European elections. On a scientific level, the unity of Europe is already a daily fact of life, and has been so for quite some time. A paper submitted to a journal is likely to be reviewed by several colleagues in Europe and elsewhere. Scientists know and visit each other. Information circulates fast and freely between us. National boundaries do not slow the diffusion of knowledge. After completing their theses, at around age 27, our researchers often go abroad to work for at least a year.

European governments promote more precocious exchanges involving students in the early stages of their education. I am not convinced that interactions between very young European students is as beneficial as one might imagine, at least not in the so-called hard sciences. As far as business schools are concerned, the approach appears to work well. But in our disciplines, younger students lack the maturity required to absorb what is necessary, filter out what is not, and adapt to an unfamiliar environment.

CHAPTER 4

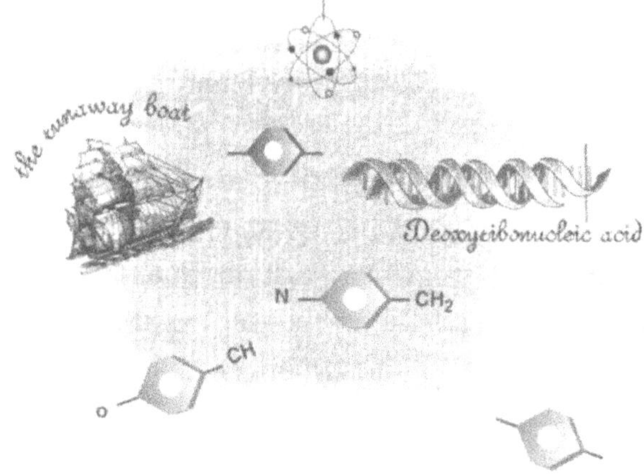

New Banners

Quite naturally, the debate surrounding education raises questions that go much beyond the pedagogical framework. A major problem in Western societies is that the objectives presented to our youth are mediocre, centered as they are on comfort and leisure. A primary duty is to define societal issues that educe a collective yearning. I perceive two such issues as being of fundamental importance: the first is *the ethics of knowledge* (once advocated by Jacques Monod), and the second is *global solidarity*. It is clear that, should we fail to persuade our young generations to take up the banner of solidarity, we are heading toward a worldwide conflict within the next 30 years, and all will be lost.

ETHICS AND SOLIDARITY

The two issues—ethics and solidarity—have common points. Without a spirit of fellowship in research, technology, and education, the great goal of North-South solidarity will remain a pious dream. This brings me to a problem that I have omitted in my discussion of the weaknesses of our universities: they teach a curriculum which is totally ill-adapted to developing countries—LDCs (least developed countries) in United Nations parlance—although they admit a large contingent of students who come from such countries. We give these students a sophisticated education which does not at all fit what they need in order to contribute to the progress of their native country.

I will use a concrete example. There was some time ago a well-intentioned French effort to introduce in Peru the technique of nuclear resonance, a delicate and beautiful spectroscopic tool, but one that is highly specialized and requires complicated hardware, difficult to operate in a Latin American country. There were problems with the Customs Bureau, with spare parts, with the lack of trained maintenance technicians. What Peru really needed was probably quite different, more along the line of technical training related to mineral chemistry, aimed at enabling the country to generate income from its national resource: the exploitation of its ores.

Many students from developing countries come to us and learn technologies that they will not be able to apply back home, and they routinely elect to remain in France. Those who do return to their native land often try to transplant forms of research or technologies that are not adapted to a developing economy. In this arena, the education system must be rethought from the ground up—and not just in France, but in all developed nations. It is one of the necessary conditions for the realization of a common North–South project, of a federating effort which would also be a message of hope about our will to survive together.

Scientific exchanges, through the bridges they kept open despite the tensions of the cold war, despite the disparities between the world's economies, despite the varied cultures and religions, can provide an excellent vehicle for this collective drive. Scientific interactions promote contacts between people seemingly in conflict over everything else. It is one of the few cements at our disposal to build a spirit of solidarity.

What should our schools cultivate in our engineers? Should they nurture the imagination, the invention of new processes which will permit old and developed countries like France to retain some degree of economic lead? This is a legitimate vision, justified by our own shortcomings and our problems of poverty, but it is still a somewhat shortsighted perspective, not nearly inspiring enough to elicit an enthusiasm comparable to that of the Renaissance.

For lack of motivating endeavors, our societies shrivel and degenerate into nationalistic, ideological, or religious sectarianism. A certain faction of the ecological movement turns out to be religious in nature, by virtue of its manichaeism (here is Good, there is Evil), its inquisitorial bent, and even its internal heresies. Our societies are hard-pressed to find a banner that is more than an ordinary rag.

An important observation has already been made, but it must be repeated here. It will serve no purpose to solve our environmental problems if we fail to address the birth rate dilemma. The demographic issue is the most difficult challenge faced by the younger generation. To ignore this problem is tantamount to accepting the stabilization of the world's population by means of massacres and famines.

"We, Civilizations, Now Know That We Are Mortal"[1]

I am not at all certain, and I use here the language of a physicist, that developed human societies constitute a stable form of life.

With the technology of present-day radio astronomy, and given the short duration of the search for extraterrestrial intelligence (only about 30 years), no signal has ever been detected providing us with a hint of the existence of organized life anywhere else in the universe. Why is that?

There is increasing indirect evidence of the existence of numerous planetary systems in the universe. There is, therefore, a nonnegligible probability that in some of these systems the right chemical conditions (soft matter, perhaps), conducive to the evolution of a form of life not too different from ours, might have prevailed. Yet we have no indication of this. Why? Life on Earth is based on a genetic code that has existed for several billion years. Yet we have had the know-how to broad-

[1] P. Valéry, *La Crise de l'Esprit* (*Crisis of the Spirit*), 1919.

cast radio signals over vast distances for a mere century. One hundred years compared to a billion represents only one part in ten million of our history. The time over which we have had the technology to emit signals is extraordinary brief.

The real question is whether our civilization is a durable phenomenon, or whether it is intrinsically unstable. If it is durable and we continue to broadcast over an appreciable fraction of a billion years, the same ought to be true elsewhere. Why, then, have we failed to detect any radio signals from alien civilizations based on other planetary systems?

It is possible that an advanced civilization, capable of radio transmission, constitutes an intrinsically unstable phenomenon. It may be that a certain level of knowledge, reflected by a mastery of communications technology, is predestined to create internal conflicts that are insoluble and self-destructive. If so, this tiny window of time—a hundred years out of a billion—might represent the only opportunity to emit signals. Should other life forms be similarly ephemeral, the chance of detecting them would be extremely low indeed.

Will the ideal of knowledge promoted by Jacques Monod (an ideal which remains difficult to inculcate in an entire society) and the ideal of global solidarity be sufficient for us to evolve into a civilization that is durable? The answer is to be written by our children.

EPILOG

While rereading these pages, a nostalgic feeling wells up in me. Feelings of regrets for having brought this period to a close. Regrets for having said too little; for having given answers which often missed the gist of the question. Most of all, regrets for having inadequately conveyed our pains and our joys as researchers—the long months, even years, when the experiment refuses to work, when everything fails; or the battle with a theoretical description which stubbornly resists attempts at simplification, streamlining, and broadening its reach, a battle which reminds me of the laborious hatching of a chrysalis. And also the doubts: Is this problem truly important? Should our friends be encouraged to work on it?

And then, from time to time, exhilarating moments when everything falls in place. When a wall of mystery crumbles to reveal a new landscape, not necessarily the Annapurna, but perhaps a gentle rolling hill in Tuscany that no painter has yet captured.

The sense of happiness when a physicist, whom one once knew as a novice, walks on his own and climbs faster than his guide! Or when a disillusioned colleague suddenly finds a rekindled enthusiasm while working on a new problem, ideally matched to his background.

Hendrik Casimir, the great master of Dutch physics, once told me: "Almost everything we do is written in sand and fades away in the wind. Once in a while, we are fortunate enough to be given a metal tablet on which to inscribe a more permanent message."

This book is written in sand. But the beach is magnificent, and I am immensely grateful to have been able to stroll along the shore.

GLOSSARY

1 POLYMER, MACROMOLECULE (long-chained molecule)

A molecule of polymer (from the Greek word meaning "in several parts"), or macromolecule (from the Greek *makro*, meaning "long," or "large"), is a molecule formed by the sequential linking of groups of atoms. The number of these groups can be large (1000, 10,000, . . .) One example is polyethylene, whose basic structure is $\cdots$ -CH_2-CH_2-CH_2-CH_2-CH_2-CH_2-$\cdots$.

2 MOLECULES

A few atoms exist in free form (for example, the inert gases like argon and krypton), but most atoms attract each other to form stable structures known as molecules. One can distinguish two kinds of molecules:

- Molecules formed by identical atoms, for instance hydrogen gas, which is denoted H_2.
- Molecules consisting of different atoms, for instance water, denoted H_2O (two hydrogen atoms and one oxygen atom).

3 ATOMS

Matter is composed of elements (or "bricks"), namely atoms. There are only 92 different elements (or 92 kinds of atoms) in nature. For simplicity, one can think of these atoms as spheres. The radius of these spheres is exceedingly small. It is of the order of 1 to 2 Å, except for hydrogen, in which case it is 0.4 Å (see [4]).

4 ANGSTROM

The *angstrom* (Å) is a unit of length often used in microphysics. It takes 10 billion Å to cover 1 meter. The size of atoms is of the order of 1 Å. The *nanometer* (nm) is one billionth of a meter (1 billion = 1000 million). The *micron* or *micrometer* (μm) is one thousandth of a millimeter, or one millionth of a meter.

5 CONDUCTORS, INSULATORS, SEMICONDUCTORS

Ordinary solids can be classified according to the magnitude of the resistance that they offer to the flow of an electric current.

Metals easily conduct electricity. Their resistance decreases as their temperature is lowered. Examples are copper and aluminum.

Insulators, also called *dielectrics*, are practically incapable of conducting electricity. Examples are glass and oil. A third category of solids, called *semiconductors*, conducts electricity weakly. The resistance of semiconductors depends strongly on their chemical composition. A classic example is silicon, which conducts rather poorly unless it contains a small concentration of impurities like boron or phosphorus. In contrast to that of metals, the resistance of semiconductors decreases with increasing temperature. The properties of semiconductors form the basis of modern electronics.

6 SUPERCONDUCTORS, SUPERCONDUCTIVITY

When the temperature of a metal is lowered, its resistance generally decreases slowly. In some cases (like lead, niobium-titanium alloy, etc.), the resistance drops to zero below a certain critical temperature. The material has become a superconductor.

7 PERIODIC TABLE (or periodic classification) OF THE ELEMENTS

During the nineteenth century, it became possible to "weigh" atoms. The ratio of the mass of the atom of an element to that of a hydrogen atom is called the *atomic mass* of that element (the hydrogen atom is the lightest of all). For example, oxygen has an atomic mass of 16.

During the same period, it was recognized that certain atoms had similar physicochemical properties. They form a kind of family.

A Russian chemist, D.I. Mendeleyev (1834–1907), had the idea of constructing a table of eight columns, containing all the elements known at the time. Each column corresponds to a family. In each column and each row of the table, Mendeleyev placed an individual element in order of increasing atomic mass. This table, which initially had many missing entries, gradually filled as new atoms were discovered which fit right into the available empty slots. The table proved to have considerable predictive powers. In the 1930s, quantum mechanics provided an understanding of this classification scheme, based on successive shells of electrons around the nucleus.

8 FISSION

When a uranium atom, with atomic mass 235, is bombarded by a slow neutron with little energy, it is split. (A neutron is one of the *elementary particles* which constitute the nuclei of atoms.) The reaction releases two lighter atoms, three neutrons, and some excess energy. The neutrons, when slowed down considerably, can, in turn, split other uranium atoms, which release more energy. This constitutes a chain reaction. The excess energy can be recovered to power turbines. This power-conversion scheme is exploited in nuclear power plants.

9 FUSION

If one manages to fuse two light atoms (for example two hydrogen atoms) into a single, heavier atom, the reaction is accompanied by the release of a vast amount of excess energy. This reaction forms the basis of the hydrogen bomb. An effort is under way to try to control it and use it as a potential source of energy.

INDEX

www.ingramcontent.com/pod-product-compliance
Lightning Source LLC
LaVergne TN
LVHW052323200926
848481LV00002B/653
* 9 7 8 1 4 6 1 2 7 5 2 8 2 *